ROYAL AGRIC

STP 1078

Pesticide Formulations and Application Systems: 10th Volume

L. E. Bode, J. L. Hazen, and D. G. Chasin, editors

ASTM
1916 Race Street
Philadelphia, PA 19103

ASTM Publication Code Number (PCN): 04-010780-48
ISBN: 0-8031-1388-9 ⁄
ISSN: 1040-1695

NOTE

The Society is not responsible, as a body,
for the statements and opinions
advanced in this publication.

Peer Review Policy

Each paper published in this volume was evaluated by three peer reviewers. The authors addressed all of the reviewers' comments to the satisfaction of both the technical editor(s) and the ASTM Committee on Publications.

The quality of the papers in this publication reflects not only the obvious efforts of the authors and the technical editor(s), but also the work of these peer reviewers. The ASTM Committee on Publications acknowledges with appreciation their dedication and contribution of time and effort on behalf of ASTM.

Printed in Ann Arbor, MI
October 1990

Foreword

This publication, *Pesticide Formulations and Application Systems: 10th Volume,* contains papers presented at the Tenth Symposium on Pesticide Formulations and Application Systems held in Denver, Colorado, on 25–26 October 1989. The symposium was sponsored by ASTM Committee E-35 on Pesticides and by its subcommittee E35-22 on Pesticide Formulations and Application Systems. L. E. Bode, University of Illinois, served as symposium chairman, and J. L. Hazen, BASF Corporation, and D. G. Chasin, ICI Specialty Chemicals, served as symposium cochairmen. These men served as coeditors of this publication.

Contents

Overview

The purpose of this tenth symposium on pesticide formulations and application systems was to continue and build on the exchange of information and research data among pesticide formulators and application scientists that has been established by the nine previous symposia (STP 764, 795, 828, 875, 915, 943, 968, 980, and 1036). The aim was to:

1. Provide a forum for exchange of ideas and data among chemists and engineers working to improve the efficiency of pesticide use.
2. Provide a data base to support ASTM Committee E35 in development of guides and standards.
3. Serve as a guide to Subcommittee E35.22 members in their future efforts to address the issues related to the use of pesticides.

This volume, in addition to previous symposia proceedings, adds significantly to the available resources on the important subject of pesticides. Subjects in this STP include the technical aspects of pesticide application and formulation research, including equipment and concepts contributing to the effective and responsible use of pesticides. Regulatory aspects of pesticides are included as an integral part of pesticide development and use. Direction and suggestions for the development of several standards were made available from material in this volume.

The 21 papers in this STP are organized into three sections. The first section, *Perspectives on Pesticide Risks,* includes two papers from the keynote session regarding risk assessment and how risk assessment imparts the regulatory actions of pesticide registration. Papers from two other keynote speakers are not available in this volume: R. L. Denny, USEPA, discussed the regulatory policies of how formulations impact on waste minimization and disposal. J. A. Graham, Monsanto Company, presented insight on how modern analytical chemistry impacts on perceptions of pesticide risks.

The second section, *Formulation Technology and Characteristics of Uptake,* includes research regarding pesticide formulation technology and efficiency of pesticide uptake into target pests. Eleven papers on the application of pesticides comprise the final section—*Application Systems.* This section includes research on sensors and controls for precise application, off-target losses from application of pesticides, and new techniques for applying pesticides.

Perspectives on Pesticide Risks

Pesticide risk assessment is a significant component of most environmental risk management programs, both in industry and government. The papers in this section describe risk assessment issues relevant to several new regulatory initiatives of EPA. The keynote session presented an overall view of risk, including the potential human health and environment/ecological risks of using agricultural chemicals. Issues examined include pesticide risks associated with generation and disposal of hazardous waste, use of inerts in formulating pesticides, and how new analytical procedures for detecting residues on food and water are affecting perceptions of pesticide risks.

Rachman described quantitative risk assessment as a tool for regulating potentially carcinogenic chemicals. Risk assessment is a methodology for estimating the "true risk" of a toxic substance. To maximize protection of public health, regulatory attention needs to focus and reduce those human risks that are most likely to be significantly large.

The issue surrounding inerts testing has become some of the highest profit issues affecting pesticide formulations today. Curcio focused on several aspects of the regulations which have become apparent as the industry moves forward in understanding and implementing the new pesticide procedures.

Formulation Technology and Characteristics of Uptake

Wing and Dailey present recent breakthroughs in microencapsulation and cyclodextrin complexes with special emphasis on how they effect contamination of groundwater. Moechnig describes the properties of a new inert granular carrier that maintains its integrity during formulation, yet disintegrates readily when exposed to rainfall. Raymond evaluated several techniques to measure airborne dust to determine which are suitable for determining the dustiness of inert carriers. His work will serve as a basis for the development of an ASTM standard procedure for measuring airborne dust.

Characteristics of spray deposit and pesticide uptake were examined by Lutrell, Manthey, Salyani, and Sundaram. Advantages and limitations of various approaches used to estimate the relative influences of spray deposits and resulting efficacy of the pests were evaluated. Results on the affect of petroleum solvents on cuticular wax and leaf-cell permeability will be useful in selecting solvents for pesticide and adjuvant formulations. Research in this area has progressed to the point of validating in field experiments responses predicted by models of spray deposits.

Application Systems

Sensors and controls for precise application were investigated by Sudduth, Gaultney, and Leedahl. Systems are now available for on-the-go sensing of soil organic matter with the adjustment of chemical rate based on input from the sensor. Field impregnation of pesticides on granular fertilizer and application with pneumatic spreaders is being accepted as a technique for minimizing the environmental impact from pesticides.

Particle dynamics affects the deposition efficiency of spray droplets. Young, Adams, and Akesson examined the dynamics of droplet spray clouds from the time of droplet formation until deposition on the target. The formation of vortices around the spray sheet were clearly demonstrated with a proposed method for measuring drift potential.

Hall and Riley discussed the low efficiency of active chemical that actually reaches the target and present techniques for reducing the level of spray drift. Regression equations are used to predict the amount of off-target deposits during application of pesticides. Krishnan studied the effect of wind conditions and boom bounce on the uniformity of application.

Ozkan has developed a technique to inject pesticides into turfgrass in order to reduce potential contamination of people, pets, and wildlife. Appel presents a procedure to treat mounds and effectively control fire ants.

These papers confirm that the objectives of the symposium were met. This STP (in conjunction with previous symposia STPs) provides a database of information regarding pesticide formulations and application systems that will guide ASTM Committee E35.22 members in the development of necessary standards.

Loren E. Bode

University of Illinois, Urbana, IL 61801; symposium chairman and editor

James L. Hazen

BASF Corporation, Research Triangle Park, NC 27502-3528; symposium cochairman and editor

David G. Chasin

ICI Chemicals, Wilmington, DE 19897; symposium cochairman and editor

Perspectives on Pesticide Risks

Nancy J. Rachman

THE ROLE OF RISK ASSESSMENT IN THE REGULATION OF PESTICIDES

REFERENCE: Rachman, N.J. "The Role of Risk Assessment in the Regulation of Pesticides," Pesticide Formulations and Application Systems: 10th Volume, ASTM STP 1078, L.E. Bode, J.L. Hazen, and D.G. Chasin, Eds., American Society for Testing and Materials, Philadelphia, 1990.

ABSTRACT: Quantitative risk assessment is an important tool for regulating potentially carcinogenic chemicals including some pesticides. Although requiring that pesticide residue tolerances not exceed a specific dietary risk level may appear to be an effective risk management approach, it ignores significant limitations of calculated risk numbers that arise from uncertainties inherent in the risk assessment process. Determining what level of risk is to be considered negligible is a matter of policy rather than science. Different federal agencies and programs consider various levels of risk acceptable, but there appears to be some consensus that risks below 10^{-6} are negligible.

KEYWORDS: Risk assessment, true risk, uncertainty, negligible risk.

During the last few years, we have seen the regulation of pesticides continually expand into new areas of concern: for example, new sources of hazard, such as inert ingredients; new sources of exposure, such as ground water or pesticide wastes; new populations at risk, such as infants and children. The use of the word "new" of course does not imply that these hazards, exposures or risks did not previously exist, but that they were not formerly singled out for special evaluation and risk-reducing regulation.

The health risk that the public appears to fear the most has been,

Dr. Rachman is a Project Manager at ENVIRON Corporation, 4350 North Fairfax Drive, Arlington, Virginia 22203.

for about 20 years, cancer. Quantitative risk assessment for suspected or known carcinogens has become a powerful tool not only in guiding regulators, but also in shaping the worries and expectations of the public. The development of risk assessment methods has made it possible to estimate cancer risks probabilistically and thus to compare and prioritize the risks of various carcinogens and exposures. However, it can be argued that since the public (including legislators and politicians) tend to interpret risk numbers not as probabilities but as scores, that is, real descriptions of numbers of cancers, risk assessment is sometimes erroneously relied upon to provide an illusion of effective risk management.

Last year's federal legislative activity concerning food safety may be an illustrative example. H.R. 3292/S. 722 (Waxman/Kennedy) would have required that every pesticide residue tolerance meet a "negligible" dietary cancer risk standard, defined as 10^{-6} (one additional case of cancer in one million persons exposed over a 70-year lifetime). In other words, the calculated risk number is to be used as an automatic risk management trigger: if larger than 10^{-6}, the risk is significant and requires reduction.

Now to the layman this may appear to be a tough, no-nonsense approach to reducing pesticide risks, but from the professional risk assessor's point of view, the idea that every risk greater than 10^{-6} is significant for the public health is a misleading oversimplification. It ignores the fact that 10^{-6}, as a benchmark for risk significance, is a policy choice and not a scientific conclusion. It also ignores an important problem with risk numbers that arises from the risk assessment process itself. This problem is, in short, that not all 10^{-6}'s (or 10^{-7}'s or 10^{-4}'s, for that matter) are equal. The key to understanding both issues lies in appreciating the procedures of risk assessment and the uncertainties inherent in every calculated risk number.

RISK ASSESSMENT

Risk assessment is a methodology for estimating the true risk of a toxic substance. In most cases it is impossible to measure or know the true, or actual, risk. Appropriate "residue reduction" studies have generally shown that the concentrations of carcinogenic pesticides, for example, to which consumers are exposed in the food they actually eat are in a very low range. The probability of directly observing a toxic response at such low concentrations is very small; enormous numbers of human subjects would have to be exposed and measured for a test at such exposure concentrations to produce useful information about the true dietary risk of a pesticide. Risk assessment has therefore become a prominent feature in the regulation of all types of potentially hazardous substances, including both naturally occurring and man-made chemicals.

Risk assessment is the systematic, scientific evaluation of what is known, and what is not known, about a substance, in order to predict the probability of harm (risk) in a particular exposure situation. The National Academy of Sciences (NAS) has defined risk assessment as consisting of four components (NAS 1983):

1.   Hazard Identification is the evaluation of all available toxicity information on the substance in order to discover the nature of the toxic response. This step involves critically reviewing laboratory

animal and/or human epidemiological (frequently occupational) data. Usually these studies provide information about toxic responses at relatively high-level exposures. Environmental Protection Agency (EPA) guidelines specify that data from the most sensitive species and strain of animal be selected in characterizing toxicity. This conservatism is sometimes questionable, for example when it results in basing human cancer risk estimates on the incidence of tumors not known to occur in humans.

2. Dose-Response Assessment is the estimation of the probable response in the concentration range to which humans are likely to be exposed. For example, in the case of a pesticide used on food crops, anticipated residues may be in the part-per-billion (ppb) range. A consumer eating treated crops might be exposed to microgram quantities on an average day. By contrast, animal studies of the pesticide would be performed at much higher levels, frequently in the parts-per-million range (ppm). Test animals would be consuming doses about a thousand times higher than exposed humans, that is, milligram to gram quantities daily.

Two kinds of mathematical adjustments have to be made in extrapolating from animal responses at high doses to human responses at low doses: an allowance for differences between laboratory animals and humans (interspecies), and an allowance for differences in biological responses at high level and low level exposures (intraspecies). Since we don't really understand the biological bases of these differences completely, these extrapolations are based on certain scientific assumptions. EPA requires that, where a choice of assumptions is possible, the more conservative approach be selected. For example, two bases for interspecies extrapolation are recognized by toxicologists and risk assessors. One is the ratio between body weights, the other, the ratio of body surface areas. Which one is most appropriate depends on the mechanism of toxicity of the substance in question. Nevertheless, EPA requires that interspecies extrapolation be based on surface area ratios. This can dramatically affect the calculated risk. The surface area scaling factor applied to rat data produces a human risk estimate about 6 times greater than the body weight ratio; if applied to mouse data, the surface area scaling factor produces a human risk estimate 14 times higher than the body weight ratio (OSTP 1985).

The end result of a dose-response assessment for a carcinogen is an estimate of the carcinogenic "potency" or Q_1^*, the increase in cancer per unit increase in dose. Because of conservative assumptions regulatory agencies require in performing steps 1 and 2, the estimated potency is a statistical "upper bound" (95% upper confidence level). This insures that the estimated potency is much more likely to overstate the "true" potency than to understate it.

It is noteworthy that for noncarcinogens, the end result of the dose response assessment is a no-observed-effect-level (NOEL) or no-observed-adverse-effect level (NOAEL). The NOEL is the highest exposure at which <u>no</u> effects have been reported in humans or experimental animals; the NOAEL is the highest exposure at which no <u>adverse</u> effects have been reported.

3.   <u>Exposure Assessment</u> is the determination of the nature and magnitude of the anticipated human exposure. The product of this step is an estimate of the <u>dose</u>, the amount of the substance that will be available within the body to cause the toxic effect. In case of carcinogens, we calculate the lifetime average daily dose (LADD), which is the amount a 70 kg person would be exposed to daily over a lifetime of 70 years. This convention is based on current scientific thinking about cancer mechanisms, which is that there is some quantifiable, though not necessarily significant, risk of cancer at any level of exposure, and the risk increases as exposure increases. The LADD is conservative because it assumes environmental levels remain constant and exposure occurs daily all throughout life.

For noncarcinogens, however, there is thought to be a "threshold" exposure. Adverse effects occur only if the threshold is exceeded. For noncarcinogens, we therefore estimate the maximum or average daily dose (MDD or ADD) a person is likely to receive.

4.   <u>Risk Characterization</u> is the assembly of results from the previous steps into a risk estimate. For carcinogens, the risk is the product of the potency times the LADD. Because of the conservative assumptions used in estimating the potency and LADD, the estimated risk is almost certainly greater than the true risk. In fact, the estimated risk is more accurately reported as a range than as a single number. A risk reported as 1×10^{-6}, for example, may really mean "a 95% probability of being 10^{-6} or less, and possibly as low as zero."

For noncarcinogens, a risk is not calculated. The NOEL or NOAEL is divided by a safety factor to estimate an allowable exposure level called the "acceptable daily intake" (ADI) or "reference dose" (RfD). The selection of appropriate safety factor(s) is governed by scientific convention. Generally, a factor of 10 is used to account for interspecies variation, multiplied by another 10 for intraspecies differences (e.g., the possibility that some individuals are more sensitive). Additional factors are incorporated if the data have limitations (ENVIRON 1988).

Risk characterization should include a thorough description of the "uncertainties" inherent in the three previous risk assessment steps. Some generic uncertainties result from the risk assessor's having to make assumptions to cover gaps in general knowledge. (As an example, we <u>assume</u> that laboratory animals respond the same as humans would; however, this may not always be true. For example, as mentioned previously, some animals develop some tumor types that humans do not, and the significance of these tumors for predicting human risk is unclear.). This type of uncertainty is common to all risk assessments. Other kinds of uncertainties arise from particular inadequacies in the available information on the specific substance of interest and on exposure to it. In the case of pesticides, for example, we are frequently lacking adequate data on the environmental concentration (i.e., actual residues in foods as consumed) and the absorption and pharmacokinetic characteristics of the chemical.

It is the discussion of uncertainties that makes it possible to judge how good an estimate of the true risk a calculated risk number is. The uncertainty evaluation also makes it possible to identify what additional data could reduce the uncertainty and improve the risk estimate. These important exercises are generally omitted from risk assessments in regulatory situations.

From this brief overview of the risk assessment method, one can see it is impossible to judge whether any calculated level of risk from a given pesticide is significant without considering of how good an estimate of the true risk that calculation is. EPA has made a start towards evaluating the uncertainties in carcinogen risk assessments by establishing a classification scheme based on the weight of the evidence of carcinogenicity. Substances evaluated for human carcinogenicity potential are assigned alphabetical codes that indicate in a qualitative way how good the evidence of carcinogenicity is (USEPA 1986). Many scientists believe, however, that the present use of the classification scheme imposes a degree of standardization in evaluating evidence that obscures the significance of particular uncertainties in the various steps of any given risk assessment. For example, there is no category for cases in which expert opinion holds that the animal data are biologically irrelevant to predicting human risk, or that carcinogenesis is preceded by a physiological effect that has a threshold, so that cancer would not be likely if exposures were below that threshold level.

If in reducing risks from toxins we are to use the risk assessment approach to its fullest potential, there can be no substitute for informed scientific judgment on a case-by-case basis. Unfortunately, the public perception is that case-by-case consideration equates with regulatory laxity, and so the public mistrust of government and scientists exerts strong pressure for highly standardized evaluations.

<u>Negligible Risk</u>

Deciding whether or not an estimated risk is significant enough to require regulation is not a scientific matter and is not a part of the risk assessment process. Decisions about risk significance are in the realm of "risk management," which is directed by society's perceptions and values and by the language of specific laws. As a result, the levels of risk that are taken to be acceptable, significant or insignificant can and do vary. Nevertheless, even though there are no specific criteria for determining what level of risk is significant, there appears to be some consensus that risks below 10^{-6} are negligible for regulatory purposes. Rodricks et al. (1987) looked at various regulatory decisions made by three U.S. government agencies: the Food and Drug Administration (FDA), Occupational Safety and Health Administration (OSHA) and the EPA (Table 1). With respect to additives intentionally added to foods, FDA considers carcinogenic risks less than 10^{-6} as insignificant ("de minimis"). However, in 1987 a federal appeals court failed to uphold this interpretation of the Delaney Clause with respect to color additives, and the implication for other kinds of additives (such as pesticides) is presently unclear. FDA has found risks in the range of 10^{-4} acceptable in the cases of naturally-occurring carcinogenic contaminants such as aflatoxin, and a few widespread industrial pollutants such as polychlorinated biphenyls (PCBs) and dioxin, that the Agency believes are not completely avoidable in today's food supply.

Higher risks are tolerated in occupational settings. Although OSHA has not defined the level of risk it considers significant, regulatory standards for workplace carcinogens accept a risk level of 10^{-3} to 10^{-4}.

Under the Safe Drinking Water Act, EPA's Office of Drinking Water holds that no level of carcinogenic risk is safe. However, due to technological limitations, higher risks may be deemed acceptable. For example, a risk of 10^{-4} is accepted for trihalomethanes, the carcinogenic byproduct of drinking water chlorination. EPA's Office of Pesticide Programs (OPP) considers 10^{-6} as insignificant risk for carcinogenic pesticides whose residues appear on raw agricultural commodities. However, higher risks are sometimes accepted since the Federal Insecticide, Fungicide and Rodenticide Act (FIFRA) directs that risks be weighed against the benefits of pesticide use. OPP also uses the 10^{-6} standard of significance to regulate pesticide residues in processed foods; however, because these residues fall under the Delaney Clause, there is, as for FDA's de minimis approach to food additives, some uncertainty as to whether this interpretation would stand up in court, as against one that requires zero risk (i.e., absolute zero exposure). For this reason, the legislation I described earlier was proposed this year to insure the application of a negligible risk standard in setting pesticide tolerances.

Resources for regulating potential environmental hazards are limited. To maximize protection of the public health, regulatory attention should be focused on identifying and reducing those human risks that are not only likely to be significantly large, but are also likely to be true risks. In order to do this, we must emphasize the uncertainties in the risk estimates we produce. A proper appreciation of these uncertainties could lead to improvements in the test data we rely upon and the way we judge their significance in predicting human risk.

TABLE 1 --
"Significant risk" decisions in federal
regulatory agencies [1,2]

FDA	Food Contaminants	10^{-4} to 10^{-6}
OSHA	Worker Protection Standards	10^{-3}
EPA	Safe Drinking Water Act: Maximum Contaminant Level Goal (MCLG's) (carcinogens)	0
	Maximum Contaminant Level (MCL's)	(B.A.T.)[3]
	Clean Air Act: Emission standards	10^{-3} (individual risk)
	Federal Insecticide Fungicide and Rodenticide Act (FIFRA): Food crop tolerances	10^{-4} to 10^{-8}

[1]   Rodricks and Taylor (1983).

[2]   Rodricks et al. (1987).

[3]   Best available technology.

Bibliography

Elements of Toxicology and Chemical Risk Assessment Revised,
    ENVIRON Corp., Washington, DC, 1988.
National Academy of Sciences (NAS), Risk Assessment in the
    Federal Government. Managing the Process, National
    Academy Press, Washington, DC, 1983.
Rodricks, J. V., Brett, S. M., and Wrenn, G. C., "Significant
    Risk Decisions in Federal Regulatory Agencies,"
    Regulatory Toxicology and Pharmacology, Vol. 7, 1987,
    pp. 307-320.
Rodricks, J. V. and Taylor, M. R., "Application of Risk
    Assessment to Food Safety Decision-Making," Regulatory
    Toxicology and Pharmacology, 1983, pp. 275-307.
U.S. Environmental Protection Agency (USEPA), "Guidelines
    for Carcinogen Risk Assessment," Federal Register,
    51:33992-34003, 24 September 1986.
U.S. Office of Science and Technology Policy (OSTP),
    "Chemical Carcinogens: A Review of the Science and Its
    Associated Principles," Federal Register 50:10371-10442,
    14 March 1985.

Lawrence N. Curcio

RECENT PROGRESS AND ISSUES ON INERTS TESTING

REFERENCE: Curcio, Lawrence N., "Recent Progress and Issues on Inerts Testing", Pesticide Formulations and Application Systems: 10th Volume, ASTM STP 1078, L. E. Bode, J. L. Hazen, and D. G. Chasin, Eds., American Society for Testing and Materials, Philadelphia, 1990.

ABSTRACT: The issues surrounding inerts testing have become some of the highest profile issues affecting pesticide formulations today. Most agree that the Environmental Protection Agency (EPA) has been correctly concerned about the lack of toxicity data for various inert components used in pesticide formulations. In many cases this concern is justified; however, in an equal number of cases, a great deal of data exists which needs to be provided to EPA in an orderly manner. This will enable suppliers to assure formulators of a continuing supply of acceptable inert materials for their future use. The following discussion will focus on several aspects of the regulations which have become apparent as the industry moves forward in understanding and implementing the new pesticide procedures. These aspects include testing requirements, animal considerations, and state versus federal requirements.

KEYWORDS: Toxicology Pesticide Regulations, inerts, inert ingredients, solvents.

Inert Registration Issues

The first major area of concern deals with the inert registration testing requirements. Industry is concerned about the proposed testing requirements and their toxicologic suitability. The first issue under this category is the repetition of tests due to Good Laboratory Practice (GLP) requirements. In many cases toxicology tests have been adequately conducted on various inert materials, but because they have been done prior to a pre-determined date there exists the possibility of repeating the study for no other reason than that it was conducted prior to GLP requirements. This appears to be an poor utilization of industry resources.

Dr. Curcio is the Manager of Environmental Affairs with Exxon Chemical Company, Hydrocarbon Solvent. He is currently responsible for pesticide application in the environmental affairs.

The second issue is the selection of species involved in the registration procedures. In many cases EPA requires that a manufacturer conduct testing on two animal models; the rat and the dog. Toxicologically, the dog may represent an imperfect model by which to assess the potential toxicity of several of the inerts used in agricultural chemical formulations today. The rat is a much more suitable model in that it has had a long history of use in a wide variety of toxicologic tests and therefore has a well-documented data base on which to determine whether or not the potential toxicities seen in the animal model may be extrapolated to man. Associated with the use of these animals is the current concern in the U.S. over animal rights issues and the anti-vivisectionist movement which seeks to reduce the number of animals used in toxicology studies. In addition, structure-activity-relationship analysis (SAR) can provide data which may help avoid unwarranted and unnecessary animal studies.

An additional concern in industry is the lead time that would be necessary once the agricultural chemical supplier and the government come to an agreement on the proposed toxicology studies needed to support the registration of the inert materials. Industry needs adequate lead time in order to advise customers of the results. In other words, if the manufacturer is required to conduct long-term studies, the government should allow an appropriate window by which to accomplish results and discuss them with the appropriate government agencies. After this time frame is determined, the Agency needs to be aware that it will take a substantial period of time for the affected supplier to communicate the results and conclusions of the study to their customers. This will allow the formulators adequate lead time to consider reformulations if the toxicologic studies prove adverse biologic effects.

When considering discussions with agencies on the results of toxicology studies, one needs to consider whether or not interpretation of the results may require an interpretation by an independent third party peer-review system much like that which exists today in other government programs, for example, the National Toxicology Program. The use of third party individuals may be appropriate in order to conclude the actual toxicologic significance of a particular finding.

Another second major area of concern deals with state registration issues. In many cases, industry is very concerned about the redundant testing which could result. This could create a significant economic concern to suppliers if they have to conduct federally-mandated testing for registration of inerts and then, in addition, address separate state issues which may require either a repetition of the study or conducting different studies. There also exists a concern regarding the overall conclusions of these different studies. Differences in interpretation could create substantial disruption in the marketplace and impair the supplier's ability to market these products across state lines. There should be a mechanism in place whereby there is one authority which determines the appropriate level of testing. This authority could also determine what types of testing need to be done in order to assure not only the safety of the inert material, but also that there will be a singular set of data required for the continued use of the inert material in the U.S.

De Minimis Constituents Issues

A third major area of concern deals with the constituents aspects of inert testing. In many cases petroleum products, for instance, may have de minimis amounts of material present in an inert ingredient. Formulators and suppliers need to know EPA's thinking on the acceptable limits of a trace amount of material which could be present. Without any guidance, EPA in a sense creates a "ban" situation whereby a product, for instance xylene, may contain part-per-million to part-per-billion levels of benzene and thereby render it unacceptable from an EPA point of view. There needs to be consideration of several different precedents which may be appropriate in determining this de minimis amount. The Occupational Safety and Health Administration (OSHA) has their Hazard Communication Standard whereby the assessment of the potential toxicity of mixtures can be based upon the percentage of a particular material in the final product. For instance, OSHA uses a one-percent cut off for all non-carcinogenic components and 0.1% for carcinogenic components. EPA and their SARA Title III requirements embrace similar classifications.

Consortium and Proprietary Issues

The last major area of concern deals with consortium testing requirements. Suppliers produce materials on the basis of chemical uniqueness. There may be a completely appropriate consortium testing arena for commodity products, such as xylene and toluene, but not for very specific products such as xylene range aromatic solvents. This concern needs to be recognized and resolved by the EPA prior to the initiation of any consortium testing. In addition, there are also questions about the application of data compensation.

In summary, the inerts testing issue is important and necessary. Communications between the Agency and inert suppliers remain critical to the successful completion of this program. State and federal requirements should be harmonized. Finally, issues on consortium testing and data compensation still are unclear and need to be thoroughly addressed. Constructive dialogue between suppliers and the EPA will help resolve many of these concerns.

Formulation Technology and Characteristics of Uptake

Robert E. Wing, William M. Doane, and Marvin M. Schreiber

STARCH-ENCAPSULATED HERBICIDES: APPROACH TO REDUCE GROUNDWATER
CONTAMINATION

REFERENCE: Wing, R. E., Doane, W. M., and Schreiber, M. M.,
"Starch-Encapsulated Herbicides: Approach to Reduce
Groundwater Contamination," Pesticide Formulations and
Application Systems: 10th Volume, ASTM STP 1078, L. E. Bode,
J. L. Hazen, and D. G. Chasin, Eds., American Society for
Testing and Materials, Philadelphia, 1990.

ABSTRACT: Non-chemically modified cornstarch effectively
encapsulates herbicides. Aqueous gelatinization of starch
at high temperature followed by active agent addition,
drying and grinding yields encapsulated products that
release active agents slowly. The rate of release can be
controlled by varying the starch type and particle size.
Preliminary data from sand and soil columns with
encapsulated herbicides show that these products offer
potential in reducing groundwater contamination.

KEYWORDS: starch, encapsulation, herbicides, slow release,
groundwater

Several reports from this laboratory have described methods for
encapsulating herbicides and other bioactive compounds within a
starch matrix [1-7]. These methods are based on dispersing the
starch in aqueous alkali followed by chemical crosslinking reactions
after the active agent has been interspersed. Both covalent and
ionic crosslinking have been used to provide encapsulated products
that are superior in performance to non-encapsulated commercial
formulations in extending activity [8], reducing evaporative and
degradative losses [9], reducing leaching [10] and decreasing dermal
toxicity [11] of the active agent.

Drs. Wing and Doane are research chemists at Northern Regional
Research Center, Agricultural Research Service, U.S. Department of
Agriculture, Peoria, IL 61604 (U.S.A.), and Dr. Schreiber is a
research agronomist at Agricultural Research Service, U.S. Department
of Agriculture, Purdue University, West Lafayette, IN 47907 (U.S.A.)

Chemical crosslinking of starch was used previously as the means to form an insoluble encapsulating matrix to entrap the active agent. Elimination of chemicals to form the matrix was desired to reduce the cost of forming the matrix and to simplify the process of encapsulation by starch. Recently we reported [12-13] a simplified process for entrapping active agents in a starch matrix that employs no chemicals to either dissolve the starch or to crosslink it. This process involves heating a dispersion of starch in water at elevated temperature and then cooling the mixture to allow the natural process of retrogradation (self crossing) to provide the necessary insolubilization. The overall process to encapsulate herbicides then is to treat a suspension of starch in water at elevated temperature, cool the starch to the desired temperature, mix in the herbicide, allow the starch to retrograde and grind the product to the desired particle size. Herbicides encapsulated in unmodified starches of various amylose content were released at different rates in strenuous water tests. Bioassay studies [14-15] of starch-encapsulated herbicides have shown excellent controlled release activity. Recent concern over pesticides, especially alachlor, causing groundwater contamination led us to evaluate several starch-encapsulated alachlor samples for controlled release to reduce groundwater pollution. This report will present some preliminary data on sand and soil column results.

EXPERIMENTAL METHODS

a. Chemicals

Starches from CPC International, Englewood Cliffs, NJ (pearl--25% amylose), American Maize Products, Hammond, IN (waxy--0% amylose), and National Starch and Chemical Co., Bridgewater, NJ (Amylon VII--70% amylose) were used. Alachlor [2-chloro-2',6'-diethyl-N-(methoxymethyl) acetanilide (94.3%)] technical grade and Lasso 4E (54% alachlor) were supplied by Monsanto Co., St. Louis, MO and used as received.

b. Steam injection (jet) cooking

Aqueous slurries (37.5% wt/vol) of each starch (180g) were passed through a laboratory-model continuous steam injection cooker (Penick and Ford, Ltd., Cedar Rapids, IA) at 135°C (waxy and pearl) or 155°C (Amylon VII) and at a flow rate of 1.3 l/min. The amount of steam entering the cooker was used to regulate the desired dispersion temperature and the back pressure was kept constant.

c. Alachlor encapsulation

All the hot starch paste from the cooker was transferred to a Sigma blade mixer and alachlor (10.6g) was added as a solid. Following mixing at 81 rpm for 30 min. the semi-solid mass was dried on trays in a hood at 28°C overnight. The mass dried to a film and was ground into 10-20, 20-40, and >40 mesh fractions. The amount of alachlor entrapped in the fraction was determined for samples washed three times with chloroform to remove any surface alachlor.

d. Percent alachlor encapsulated

Chloroform-washed samples (0.100g) in a 1% Theramyl 120 (Nova Enzyme Corp., Wilton, CT) solution (50 ml) were shaken for 1-3 days at room temperature to dissolve the starch. The samples were extracted with chloroform and diluted to 25 ml. Alachlor concentration was measured in a Tracor 560 gas-liquid chromatograph (Austin, TX) with a Chromosorb WHP column (6 ft, 0.085 in. I.D.). A standard curve was made with trifluralin as an internal standard. The unwashed samples were also analyzed by this procedure to determine total alachlor recovered (surface and encapsulated).

e. Swellability

Samples (0.20 g, 20-40 mesh) were placed in a 10-ml graduated cylinder with water (4.0 ml) at 30°C and gently stirred several times during the first three hours to prevent clumping. After 24h the height of the swollen product in the cylinder was measured.

f. Alachlor release studies

Samples (82-215 mg) containing 5.0 mg alachlor and water (25 ml) in 125-ml Erlenmeyer flasks were agitated on an orbital shaker (Lab-Line, Melrose Park, IL) at 200 rpm. Water volume was 20% in excess of that required to dissolve all the alachlor from the product (alachlor solubility is 242 mg/l water at 25°C--Herbicide Handbook--Weed Science Soc. of America, 1974, p. 9). Experiments were terminated at various times between 15 min and 3 days by extracting the water phase with chloroform (25 ml). Alachlor concentrations were determined by GC with trifluralin as an internal standard.

g. Sand column leaching studies

Acid-washed sand (50 g) was slurried into glass columns (2.54 cm. i.d. x 30 cm.) to a height of 20 cm. On top of the wet sand in different columns was placed Lasso 4E (22 mg), 20-40 mesh pearl granules (320 mg), or 20-40 mesh Amylon granules (268 mg) to give each a loading of 10 mg alachlor. The samples were covered with a piece of filter paper and 2 cm. water. A tube from a separatory funnel reservoir remained in this water layer. Water was passed through the column at a controlled rate of 2.0 ml/min. and the effluent was collected in 10-ml fractions. The fractions were extracted with chloroform and were analyzed by GC.

h. Soil column leaching studies

Dry-screened (1/2 cm hardware cloth) Miami silt loam top soil was packed into 7.5 cm dia. aluminum tubes. Each tube had a 2.5-cm wide slot, 40-cm long down one side. This slot was covered and sealed with floral clay to produce a water-tight seal. The dry soil was scooped up and poured into the column. After each scoop was added the entire column was dropped 4 times onto a 7.5-cm dia. rubber stopper from a height of 5 cm. This procedure was repeated until the soil reached 2 cm above the top of the slot. The columns were saturated with ~750 ml water. Starch-encapsulated alachlor (20-40 mesh) and commercial Lasso EC samples were applied to the top of the columns at the rate of 3.36 kg/ha. Sand (1 cm) was placed on the samples and 250 ml water (equivalent to 5 cm rainfall) was passed through the column at a rate of 4 ml/min. A 1-cm head of water was always maintained during the leaching period to prevent channeling. After leaching the column was placed on its side and the slot was opened. Foxtail seeds (~200) were sown on the exposed dirt and a 0.5-cm layer of sand was added to cover the seeds. The sand was kept slightly moist during the 14-day growing period. The distance from the top of the slot to where the foxtail growth started was used as the leaching effect.

RESULTS

Effect of Starch Type

Several alachlor-encapsulated products were prepared with either waxy, pearl, or Amylon VII starches. Table 1 summarizes data of product formulation and analysis. As the amount of amylose in the matrix increases the amount of alachlor recovered (surface and encapsulated) decreases and the product swellability decreases. Rate of alachlor release was measured in a strenuous water test with a 20% excess of water over that required to dissolve all 5 mg alachlor in a product sample. Figure 1 shows that increasing the amount of amylose in the encapsulated matrix reduced the percent of alachlor release with time.

Effect of Particle Size

Table 1 summarizes data of the different particle sizes in the ground products. More alachlor is encapsulated in the larger particles. Figure 2 shows the rate of release of alachlor from the different size particles of Amylon VII. As the particles get smaller the rate of release increases.

Alachlor Leaching from Sand Columns

Figure 3 shows the leaching profiles of alachlor from Lasso 4E and the 20-40 mesh pearl and Amylon VII encapsulated samples. Lasso

TABLE 1 -- Effect of starch type and particle size
on encapsulation of alachlor and properties of the matrices[a]

Starch type	Mesh size	Alachlor Encapsulated[b] %	Swellability in Water[c] %
Waxy	10-20	75	disperses
Waxy	20-40	67	disperses
Waxy	>40	47	disperses
Pearl	10-20	77	260
Pearl	20-40	62	280
Pearl	>40	47	300
Amylon VII	10-20	77	180
Amylon VII	20-40	75	200
Amylon VII	>40	55	220

[a]180g starch in 300ml H_2O; 10.6g alachlor, to give total of 5%
a.i.; jet cooked at 135°C.
[b]Alachlor remaining after extracting surface alachlor from product.
[c]Sample (0.2g) in H_2O (4ml).

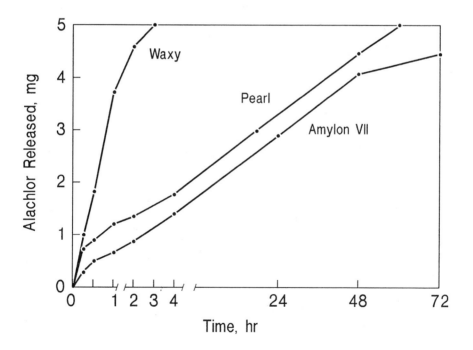

Figure 1. Effect of Starch Type of Alachlor Release

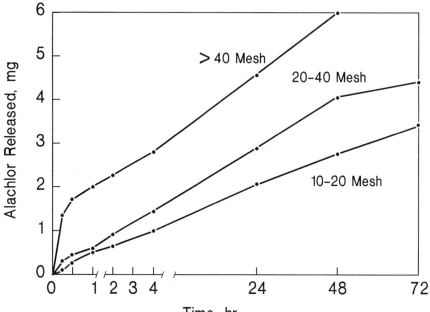

Figure 2. Effect of Particle Size on Alachlor Release

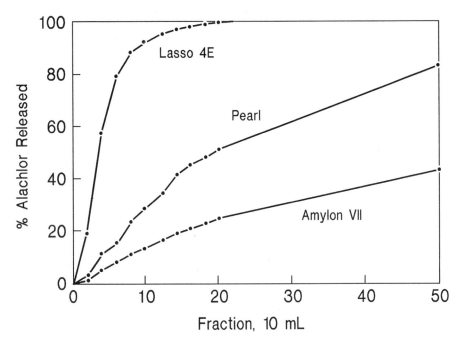

Figure 3. Sand Column Leaching of Alachlor from Starch Formulations
and Lasso 4E

4E leaches very fast as 95% of the alachlor is found in the first ten aliquots. As the amylose content increases in the starch samples the alachlor leaching decreases.

Alachlor Leaching in Soil Columns

Figure 4 shows the depth to which the alachlor leached when 250 ml water was passed through the soil columns. All starch-encapsulated samples were unwashed so the surface alachlor was still present. Again it can be seen that the starch type determines the rate of release. Other tests with different mesh sizes gave results that were consistent with previously described studies.

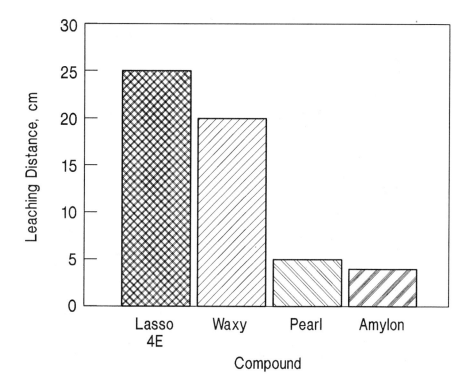

Figure 4.   Soil Column Leaching of Alachlor from Starch Formulations and Lasso 4E

CONCLUSIONS

Use of the starch matrix is effective for encapsulating herbicides like alachlor. Rate of active agent release can be controlled by the type of starch used and the particle size of the product. Preliminary results show a great reduction in leaching if the alachlor is encapsulated.

REFERENCES

[1] Shasha, B. S., Doane, W. M., and Russell, C. R., "Starch Encapsulated Pesticides for Slow Release," Journal of Polymer Science, Polymer Letters Edition, Vol. 14, 1976, pp. 417-420.

[2] Shasha, B. S., Trimnell, D., and Otey, F. H., "Encapsulation of Pesticides in a Starch-Calcium Adduct," Journal of Polymer Science, Polymer Chemical Edition, Vol. 19, 1981, pp. 1891-1899.

[3] Trimnell, D., Shasha, B. S., Wing, R. E., and Otey, F. H., "Pesticide Encapsulation Using a Starch-Borate Complex as Wall Material," Journal of Applied Polymer Science, Vol. 27, 1982, pp. 3919-3928.

[4] Wing, R. E. and Otey, F. H., "Determination of Reaction Variables for the Starch Xanthide Encapsulation of Pesticides," Journal of Polymer Science, Polymer Chemical Edition, Vol. 21, 1983, pp. 121-140.

[5] Shasha, B. S., Trimnell, D., and Otey, F. H., "Starch-Borate Complexes for EPTC Encapsulation," Journal of Applied Polymer Science, Vol. 29, 1984, pp. 67-73.

[6] Trimnell, D., Shasha, B. S., and Otey, F.H., "The Effect of α-Amylases Upon the Release of Trifluralin Encapsulated Starch," Journal of Controlled Release, Vol. 1, 1985, pp. 183-190.

[7] Wing, R. E., Maiti, S., and Doane, W. M., "Factors Affecting Release of Butylate from Calcium Ion-Modified Starch-Borate Matrices," Journal of Controlled Release, Vol. 5, 1987, pp. 79-89.

[8] Coffman, C. B. and Gentner, W. A., "Persistence of Several Controlled Release Formulations of Trifluralin in Greenhouse and Field," Weed Science, Vol. 28, 1980, pp. 21-23.

[9] Schreiber, M. M., Shasha, B. S., Rose, M. A., Orwick, P. L., and Edgecomb, D. W., "Efficacy and Rate of Release of EPTC and Butylate from Starch Encapsulated Formulations Under Greenhouse Conditions," Weed Science, Vol. 26, 1978, pp. 679-686.

[10] Baur, J. R., "Release Characteristics of Starch Xanthide Herbicide Formulations," Journal of Environmental Quality, Vol. 9, 1980, pp. 379-382.

[11] Riley, R. T., "Starch Xanthate Encapsulated Pesticides: A Preliminary Toxicological Evaluation," Journal of Agricultural Food Chemistry, Vol. 31, 1983, pp. 202-206.

[12] Wing, R. E., Maiti, S., and Doane, W. M., "Effectiveness of Jet-Cooked Pearl Cornstarch as a Controlled Release Matrix," Starch/Stärke, Vol. 39, 1987, pp. 422–425.

[13] Wing, R. E., Maiti, S., and Doane, W. M., "Amylose Content of Starch Controls the Release of Encapsulated Bioactive Agents," Journal of Controlled Release, Vol. 7, 1988, pp. 33–37.

[14] Schreiber, M. M., Wing, R. E., Shasha, B. S., and White, M. D., "Bio-Activity of Controlled Release Formulations of Herbicides in Starch Encapsulated Granules," Proceedings of International Symposium on Controlled Release of Bioactive Material, Vol. 15, 1988, pp. 223–224.

[15] Schreiber, M. M., White, M. D., Wing, R. E., Trimnell, D., and Shasha, B. S., "Bioactivity of Controlled Release Formulations of Starch Encapsulated EPTC," Journal of Controlled Release, Vol. 7, 1988, pp. 237–242.

Oliver D. Dailey, Jr., Clyde C. Dowler, and Norman C. Glaze

EVALUATION OF CYCLODEXTRIN COMPLEXES OF PESTICIDES FOR
USE IN MINIMIZATION OF GROUNDWATER CONTAMINATION

REFERENCE: Dailey, O. D., Jr., Dowler, C. C., and Glaze, N.
C., "Evaluation of Cyclodextrin Complexes of Pesticides for Use
in Minimization of Groundwater Contamination," Pesticide
Formulations and Application Systems: 10th Volume, ASTM
STP 1078, L. E. Bode, J. L. Hazen, and D. G. Chasin, Eds.,
American Society for Testing and Materials, Philadelphia, 1990.

ABSTRACT: Recently, concern over the contamination of
groundwater by pesticides has mounted. The cyclodextrin
complexes of a number of pesticides most frequently implicated
in groundwater contamination have been prepared.
    Initial pesticide efficacy studies have been conducted using
conventional application technologies with various preemergence
herbicides (as both commercial formulations and cyclodextrin
complexes) under greenhouse conditions. The leachability and
persistence of the newly formulated herbicides in the soil profile
were also measured utilizing bioassay indicator crops in the
analyses. Cyclodextrin formulations of atrazine and simazine
were prepared only under forcing conditions and had very low
levels of biological activity. A cyclodextrin formulation of
metribuzin was somewhat less active than the commercial
formulation, possibly as a result of controlled release properties.

KEYWORDS: chemigation, herbicide, atrazine, metribuzin,
simazine, ß-cyclodextrin, formulation, controlled release, Texas
panicum, Florida beggarweed, smallflower morningglory

Recently, concern over the contamination of groundwater by pesticides
has mounted. In 1986, the U. S. Environmental Protection Agency
disclosed that at least 17 pesticides used in agriculture had been found in
groundwater in 23 states [1]. According to a 1988 interim report, 74
different pesticides have been detected in the groundwater of 38 states from

---

Dr. Dailey is a research chemist at the Southern Regional Research
Center, USDA, ARS, P. O. Box 19687, New Orleans, LA 70179. Dr. Dowler
is a research agronomist and Dr. Glaze is a plant physiologist at the
Nematodes, Weeds, and Crops Research Unit, USDA, ARS, Georgia Coastal
Plain Experiment Station, P. O. Box 748, Tifton, GA 31793.

all sources. Contamination attributable to normal agricultural use has been confirmed for 46 different pesticides detected in 26 states [2]. The chief objectives of our research are to develop pesticide formulations that will maintain or increase efficacy on target organisms when applied through irrigation and that will not adversely impact on the environment or groundwater while maintaining effective pest control.

Cyclodextrins are macrocyclic torus-shaped polymers consisting of six or more D-glucose residues. They are formed by enzymatic starch degradation. ß-Cyclodextrin (BCD) is composed of seven D-glucose units connected by the 1 and 4 carbon atoms. The high electron density internal cavity (inside diameter approximately 7.5 A) consists of glycosidic oxygen bridges. Seven primary hydroxyl groups project from one outer edge of the BCD molecule, and fourteen secondary hydroxyl groups from the other. Consequently, the BCD molecule has a hydrophobic center and relatively hydrophilic outer surface. In aqueous solution, the cyclodextrin molecule readily accepts a guest molecule in its hydrophobic central cavity. It is necessary for only a portion of the molecule to fit in the cavity for an inclusion complex to form [3-4].

Many synthetic pesticides can form inclusion complexes with cyclodextrins, often resulting in improvements in the properties of the complexed substances. Cyclodextrins have found particular application for the formulation of poorly water soluble, volatile, or unstable herbicides. Among the advantages of cyclodextrin complexes of pesticides are enhanced stabilization, reduced volatility and bad odor, enhanced wettability, solubility and bioavailability, and controlled release properties. Of the cyclodextrins, BCD is the only one available at a reasonable price, and its use may be economically feasible within a few years [4]. Several herbicides that have been frequently implicated in groundwater contamination [1-2] were selected as candidates for complexation with BCD in an attempt to develop formulations that prevent entry of the chemical into the groundwater while maintaining effective weed control.

Research for the past 10 to 12 years has shown that irrigation application technology (chemigation) can be used to apply numerous pesticides in crop production systems [5]. In some cases, chemigation allows for the reduction of total chemical quantities used, thus decreasing the possibility of leaching to groundwater. The high application volumes used in chemigation generally do not decrease efficacy of soil-applied pesticides but may reduce efficacy of foliage-applied pesticides such as postemergence herbicides or insecticides by high dilution or wash off from plant foliage. Cyclodextrin complexation may facilitate retention of the pesticide on the target area during and after application [6].

Our immediate objective is to determine efficacy of the BCD complexes of selected herbicides (atrazine, metribuzin, and simazine) on the appropriate indicator crop and selected weed species as compared to commercial formulations of these herbicides and to measure biologically the potential release rate of the complexes versus commercial formulations. The ultimate objective is to measure and utilize these BCD complex formulations in conventional and chemigation applications to reduce potential groundwater contamination.

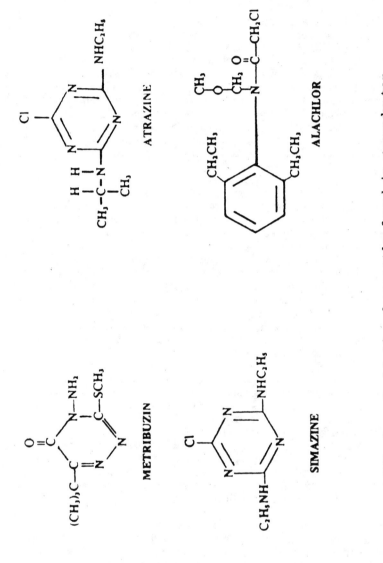

FIG. 1 -- Herbicides frequently found in groundwater

EXPERIMENTAL METHOD

Preparation of BCD Complexes

The following herbicides frequently implicated in groundwater contamination were selected as candidates for complexation with BCD: atrazine, simazine, metribuzin, and alachlor (Fig. 1). Typical reaction conditions for the formation of BCD complexes of these herbicides are shown in Table 1.

TABLE 1 -- Cyclodextrin complexation of selected herbicides

Herbicide	Complex Formation?	Reaction Conditions (Aqueous solution)	Analysis[a]
Metribuzin	Yes	65-75°C, 20 min.-6 h	1:1:5
Atrazine	No	60°C, 45 min.	...
Atrazine	Yes	100°C, 7-18 days	1:1:5
Alachlor	No	100°C, 7 days	...
Simazine	Yes	100°C, 11 days; 100°C, 2 days after adding 10% (v/v) dioxane	2:1:5

[a]Herbicide:BCD:water molar ratio as determined by elemental analysis.

The BCD formulations of atrazine were prepared by adding technical grade atrazine to a solution of an equimolar amount of BCD in water in a round-bottom flask equipped with condenser at 60-80 °C and then refluxing under an argon atmosphere until all solid had dissolved. The reaction mixtures were allowed to cool to room temperature, and any precipitate was filtered. The resulting solutions were utilized in the greenhouse studies reported below. The solid BCD complex was isolated from a portion of each solution (by removal of water at 1-2 torr and 30-40 °C) for characterization purposes. Reaction times varied, with longer periods of time required for more concentrated solutions. In the first preparation, a mixture of 10.0 g (46.4 mmol) of atrazine and 56.8 g (46.4 mmol) of BCD was heated at 100 °C in 700 ml of water for 18 days. Formulation BCD1 was obtained as an orange solution of 63.71 g of complex in 500 ml of water (0.0884 M, equivalent to 9.54 g of atrazine). The orange coloration was evidence of the presence of decomposition products formed during the extended reaction time. Indeed, the NMR spectrum of the isolated solid product indicated partial decomposition of the BCD. In the second and third preparations, 5.00 g of atrazine and 28.4 g of BCD were heated at 100 °C in 500 ml of water for 8 and 9 days, respectively. Formulation BCD2 was obtained as a colorless solution of 26.95 g of complex in 500 ml of water (0.0374 M, equivalent to 4.03 g of atrazine). Formulation BCD3 was obtained as a colorless solution of 29.15 g of complex in 500 ml of water (0.0404 M, 4.36 g of atrazine).

In the preparation of the BCD formulation of metribuzin, a solution of 5.00 g (23.3 mmol) of metribuzin in 15 ml of methanol (60 °C) was added to a solution of 28.58 g (23.3 mmol) of BCD at 70 °C. The mixture was heated at 70-75 °C under argon for 5.75 hours, at which time all solids had dissolved. After cooling, a small amount of precipitate was filtered, and 491 ml of solution containing 27.90 g of complex (0.0395 M, equivalent to 4.15 g of metribuzin) was isolated.

In the preparation of the BCD formulation of simazine, a mixture of 5.00 g (24.8 mmol) of technical grade simazine and 30.4 g (24.8 mmol) of BCD in 500 ml of water was heated at 100 °C for 11 days. There was no significant dissolution of the simazine. Fifty ml of 1,4-dioxane was added as a co-solvent, and the mixture heated at 100 °C for 2 days, yielding a clear solution. After cooling to room temperature 392 mg of white precipitate (unreacted simazine) was isolated; after refrigeration of the filtrate at 4 °C for 5 days, 12.99 g of unreacted BCD came out of solution. Removal of solvent from the remaining solution followed by vacuum drying provided 22.12 g of BCD complex. Elementary analysis (Table 1) of the material indicated a 2:1:5 simazine/BCD/H_2O molar ratio (24.8% simazine). All of the complexes contained 5 molecules of water per molecule of BCD. Elemental analysis of the BCD used in the complex preparations showed a 5:1 water/BCD ratio.

All of the BCD complexes were characterized by physical properties, NMR, IR, and UV spectra. These measurements established the formation of true inclusion compounds as opposed to mechanical mixtures of herbicide and BCD and will be presented in detail in a future paper.

Greenhouse studies: A 10-10-10 fertilizer was thoroughly mixed at the rate of 1,000 kg/ha into an air-dried Tifton loamy sand top soil and placed in 20 X 35 cm galvanized steel flats. The soil was then uniformly moistened by sprinkler from the top and allowed to equilibrate for 24 hours. The appropriate indicator crop of soybeans for metribuzin and corn for atrazine and simazine and the selected weed species were then planted in rows 3 cm apart in individual flats. The flats were again lightly moistened with overhead sprinklers and the herbicides applied preemergence to crops and weeds. The applications were made with an enclosed chamber greenhouse sprayer using a Tee Jet 80067 flat fan spray tip operating at 160 kPa which delivered a volume of 187 l/ha at 0.45 m/sec. Spray height was 46 cm. The treated flats were placed in a greenhouse with day length maintained at approximately 14 hours and the temperature ranging from 20 to 34 °C.

In a preliminary experiment with cyclodextrin formulations of atrazine (Table 2) percent control of selected weeds was recorded three weeks after treatment based on 0 = no effect and 100 = complete kill, as compared to an untreated check. In subsequent experiments with atrazine and metribuzin (Table 3), efficacy or percent control was measured by counting emerged plants and calculating percent control as compared to the untreated check two weeks after planting. The flats were then allowed to air dry, the tops of dead plants carefully removed without disturbing the soil, and the same crops or weeds replanted 3.5 and 7 weeks after initial treatment to get an estimate of herbicide persistence or release. In a third experiment with BCD complexes of atrazine, simazine, and metribuzin (Table 4) percent control was recorded two weeks after treatment. In all of the above

TABLE 2 -- Efficacy of β-cyclodextrin formulations of atrazine on selected weed species.

Herbicide[b]	Formulation	Percent Control[a]				
		Texas panicum	Smallflower morningglory	Florida beggarweed	Florida pusley
atrazine	BCD1	2	36	23	30
atrazine	BCD2	0	10	30	17
atrazine	90 DF	87	100	98	98

[a]Visual comparison to untreated check 3 weeks after treatment.

[b]Applied preemergence at 1.5 kg/ha.

TABLE 3 -- Percent control of selected weed species with β-cyclodextrin formulations of metribuzin and atrazine.[a]

Herbicide	Formulation	Rate kg/ha	Texas panicum			Smallflower morningglory			Florida beggarweed			Palmer amaranth		
			0[b]	3.5	7	0	3.5	7	0	3.5	7	0	3.5	7
metribuzin	BCD	0.4	38	0	0	100	37	18	100	86	11	100	100	0
metribuzin	75 DF	0.4	76	40	18	100	63	27	100	100	44	100	100	0
atrazine	BCD2	1.6	0	0	27	13	0	0	0	0	0	0	0	0
atrazine	90 DF	1.6	46	0	0	100	100	100	100	100	100	100	100	100

[a]Based on weed emergence in untreated check.

[b]Date of planting after treatment (weeks).

TABLE 4 -- Percent control of selected weed species with β-cyclodextrin formulations of atrazine, simazine, and metribuzin applied preemergence.[a]

Herbicide	Formulation	Rate kg/ha	Texas panicum	Smallflower morningglory	Florida beggarweed	Palmer amaranth
atrazine	BCD3	1.6	10	0	0	0
atrazine	90DF	1.6	12	100	100	100
simazine	BCD	1.7	40	0	0	0
simazine	80 W	1.7	2	100	100	100
metribuzin	BCD	0.4	38	100	100	98
metribuzin	75 DF	0.4	53	100	96	100

[a]Treated 8/2/89, counted 8/16/89, based on weed emergence in untreated check.

experiments, conventional herbicide treatments were made for comparison.

The weed species included in these experiments were Texas panicum (Panicum texanum Buckl.), smallflower morningglory [Jacquemontia tamnifolia (L.) Griseb.], Florida beggarweed [Desmodium tortuosum (Sw.) DC], Florida pusley (Richardia scabra L.), and Palmer amaranth (Amaranthus palmeri S. Wats). These species were chosen for their extensive occurrence in the Southeastern United States and for their different levels of resistance to the herbicides used in these experiments.

Procedure on leaching studies: To determine leaching characteristics and potential for groundwater contamination of slow release herbicides, the following procedure was developed. Columns 10 cm in diameter and 46 cm long were constructed of PVC pipe with a removable cover the length of the column so that cucumbers as a bioassay plant could be planted after treatment and leaching were completed. This bioassay procedure delineates the concentration and movement of herbicide throughout the column.

After construction the columns were filled with a Lakeland sand. The columns were then moistened with sufficient water to wet the entire 46 cm depth. After equilibration for 72 hours, herbicide treatments were applied by conventional techniques as described earlier. After 24 hours, 800 ml of water were added to each column, which was sufficient to cause leaching from the bottom of each column. After drainage ceased, the covers were removed from the columns and cucumbers planted the length of the columns. Germination and growth were measured 14 days later.

Results and Discussion

In the preparation of the BCD complexes, there was excellent correlation of the experimental results with computer molecular model studies [7]. These studies indicated that the entire metribuzin molecule can be accommodated inside the BCD central cavity with relative ease. Indeed, a metribuzin-BCD complex was formed under mild conditions. Model studies of the atrazine-BCD complex indicated that at least part of the atrazine molecule and probably the entire molecule (with some crowding) can fit inside the BCD central cavity. One would predict formation of an atrazine-BCD complex, but with greater difficulty than the metribuzin complex. Experimentally, atrazine was recovered unchanged under mild reaction conditions (60 °C, 45 min.) but a complex was formed under more forcing conditions (100 °C, 7 days). Model studies of the alachlor-BCD complex indicated that at best only a small portion of the alachlor molecule can fit inside the BCD central cavity. Experimentally, no BCD complex was formed under the most forcing conditions (100 °C, 7 days).

In general, BCD complexes are less soluble in water than either BCD or the guest molecule. The BCD complexes of atrazine, metribuzin, and simazine, however, were considerably more soluble in water. The solubility of BCD in water at 25 °C is 1.85 g/100 ml [8,9]. The water solubilities of the herbicides at 20 °C are as follows [10]: atrazine, 28 mg/l; metribuzin, 1.2 g/l; simazine, 8.5 mg/l. The water solubilities of the BCD complexes of the herbicides were determined at 21 °C and are as follows: atrazine, 9.29 g/100 ml; metribuzin, 4.13 g/100 ml; simazine, 5.32 g/100 ml.

The BCD formulations of atrazine (BCD1 and BCD2) did not control Texas panicum, smallflower morningglory, Florida beggarweed, and Florida pusley when compared to the commercial formulation of atrazine (Table 2). Smallflower morningglory, Florida beggarweed, and Florida pusley are highly susceptible to atrazine and are normally controlled at rates below those applied in this test. Acceptable control of Texas panicum must exceed 90% [11], while 75% would be acceptable for moderate infestations of smallflower morningglory, Florida beggarweed, Florida pusley, and Palmer amaranth. Field corn was not injured by the herbicide treatments.

In a second set of experiments, BCD formulations of metribuzin and atrazine (BCD2) were compared to commercial formulations for soybeans or corn and the weeds Texas panicum, smallflower morningglory, Florida beggarweed, and Palmer amaranth for initial efficacy and potential controlled release over a period of 7 weeks. The results are shown in Table 3. The BCD complex of metribuzin had a low level of activity on Texas panicum after treatment but controlled smallflower morningglory, Florida beggarweed, and Palmer amaranth. The commercial formulation of metribuzin (75 DF) resulted in moderate control of Texas panicum after treatment and controlled the other weed species. The activity of the metribuzin-BCD complex decreased on Texas panicum, smallflower morningglory, and Florida beggarweed at 3.5 and 7 weeks after treatment. The commercial formulation of metribuzin also resulted in decreased activity on these three weed species at 3.5 and 7 weeks after treatment, but the activity was always higher than that of the BCD formulation. Texas panicum and smallflower morningglory have some resistance to metribuzin over time. The results in Table 3 show that Florida beggarweed and Palmer amaranth are extremely susceptible to metribuzin. However, at 7 weeks after treatment, the activity decreased dramatically. The results show that the BCD formulation of metribuzin may give initial efficacy equivalent to the commercial formulation, but that the amount of metribuzin remaining in the soil 3.5 or 7 weeks after treatment is generally less than the commercial formulation. This indicates that the amount of active metribuzin in the soil is less with the cyclodextrin formulation than with the commercial formulation, possibly as a result of controlled release properties. The results on Texas panicum would indicate that the initial concentration of metribuzin in the BCD formulation is not as high as the commmercial formulation, and this difference persists for at least 7 weeks after planting. Soybeans were not injured by the metribuzin treatments.

The cyclodextrin formulation of atrazine (BCD2) had little or no activity as compared to the commercial (90 DF) formulation of atrazine (Table 3). The commercial formulation controlled smallflower morningglory, Florida beggarweed, and Palmer amaranth at 0, 3.5, and 7 weeks after treatment. Atrazine did not control Texas panicum. The overall results would indicate that atrazine is tied up in the BCD complex and is not available in active form in the soil. Corn was not injured by the atrazine treatments.

In the two-week study summarized in Table 4, all formulations of atrazine, simazine, and metribuzin (applied preemergence) gave poor and erratic control of Texas panicum. This is consistent with previous results. The cyclodextrin formulations of atrazine (BCD3) and simazine were biologically inactive on smallflower morningglory, Florida beggarweed, or Palmer amaranth. The commercial formulations controlled these species. Smallflower morningglory, Florida beggarweed, and Palmer amaranth were controlled by the BCD and commercial formulations of metribuzin. Corn was not injured by atrazine or simazine and soybeans were not injured by metribuzin.

In the leaching studies, the commercial formulation of atrazine killed 34 % of the cucumbers to a depth of 21 cm. The cyclodextrin complex of atrazine (BCD2) remained biologically inactive as indicated in the greenhouse efficacy studies. The biological results indicate that no detectable free atrazine was leached from the BCD complex but do not preclude the possibility that the BCD complex itself has leached. A method for detection of BCD complexes by HPLC is being developed for future leaching studies.

Conclusions

ß-Cyclodextrin complexes of atrazine and simazine were prepared only under forcing reaction conditions over lengthy periods of time. The greenhouse studies presented in this paper show that the formulations of these complexes are ineffective in the control of weeds. Both atrazine and simazine have very low solubilities in water. Apparently, these compounds must first dissolve in water before complexing with BCD, accounting for the long reaction times. Once formed, the BCD complexes of atrazine and simazine are highly stable and impervious to disassociation. Metribuzin forms a complex with BCD with relative ease under mild reaction conditions. The metribuzin-BCD formulation gave promising results in the greenhouse efficacy studies and will be evaluated further. In conclusion, there appears to be a direct correlation between the ease of preparation of the BCD complex of a herbicide and its biological activity. The more difficult it is to prepare a complex, the less herbicidal activity it exhibits.

ACKNOWLEDGMENTS

The authors thank James V. Kelly for technical assistance. We also acknowledge with appreciation American-Maize Products Company, Hammond, Indiana, for supplying samples of beta-cyclodextrin, Ciba-Geigy, Greensboro, North Carolina for samples of atrazine and simazine, Mobay Chemical Corp., Kansas City, Missouri, for samples of metribuzin, and Monsanto, St. Louis, Missouri, for a sample of alachlor.

Mention of a trademark, proprietary product or vendor does not constitute a guarantee or warranty of the product by the U. S. Department of Agriculture and does not imply its approval to the exclusion of other products or vendors that may also be suitable.

REFERENCES

[1] Cohen, S. Z., Eiden, C., and Lorber, M. N., "Monitoring Ground Water for Pesticides," in Evaluation of Pesticides in Ground Water, Garner, W. Y., Honeycutt, R. C., and Nigg, H. N., Eds., ACS Symposium Series, No. 315, American Chemical Society, Washington, D. C., 1986,

pp 170-196.

[2] Williams, W. M., Holden, P. W., Parsons, D. W., and Lorber, M. N., "Pesticides in Ground Water Data Base: 1988 Interim Report," U.S.E.P.A., Office of Pesticide Programs, Environmental Fate and Effects Division, Washington, D. C., 1988.

[3] Pagington, J. S., "ß-Cyclodextrin: the Success of Molecular Inclusion," Chemistry in Britain, Vol. 23, May, 1987, pp 455-458.

[4] Szejtli, J., "Cyclodextrins in Pesticides," Starch, Vol. 37, No. 11, 1985, pp 382-386.

[5] Johnson, A. W., Young, J. R., Threadgill, E. D., Dowler, C. C., and Sumner, D. R., "Chemigation for Crop Production," Plant Disease, Vol. 70, No. 11, 1986, pp 998-1004.

[6] Young, J. R. and Sumner, D. R., Eds., Second National Symposium on Chemigation, Tifton, Georgia, 1982.

[7] Dailey, O. D. and French, A. D., unpublished results.

[8] Pszczola, D. E. Food Technology (Chicago), Vol. 42, No. 1, 1988, pp. 96-100.

[9] Saenger, W. Angewante Chemie, International Edition in English, Vol. 19, 1980, pp. 344-362.

[10] Hartley, D. and Kidd, H., Eds. The Agrochemicals Handbook, Second Edition, The Royal Society of Chemistry, The University, Nottingham, England, 1987.

[11] Dowler, C. C. and Hauser, E. W., "Texas Panicum Control in Corn," Southern Weed Science Science Society Proceedings, Vol. 24, 1971, pp. 148-155.

Bruce W. Moechnig and Burnie L. Wilhelm

DEVELOPMENT AND PROPERTIES OF TRANSPORT -- A NEW INERT
GRANULAR CARRIER

REFERENCE: Moechnig, B. W. and Wilhelm, B. L., "Develop-
ment and Properties of TRANSPORT -- a New Inert Granular
Carrier", Pesticide Formulations and Applications Systems:
10th Volume, ASTM STP 1078, L. E. Bode, J. L. Hazen and
D. G. Chasin, Eds., American Society for Testing and
Materials, Philadelphia, 1990.

ABSTRACT: A new inert granular carrier has been developed
by Cargill, Inc. for use in the formulation of granular
insecticide/herbicide/fungicide/fertilizer products. The
carrier is made by agglomerating finely divided organic
material using a suitable binder, followed by drying and
screening to the desired final moisture content and particle
size. Physical property measurements show a bulk density
in the range of 0.48-0.56 g/cc, water absorption of 29.9-
47.5% and resistance to attrition of greater than 95%.
Moisture content and particle size distribution can be
varied by adjustments to the process. Comments from
potential users indicate that this new carrier maintains its
integrity during formulation, yet our results show that it
disintegrates readily when placed in water. As a result,
products formulated with this carrier have the potential
for reduced avian toxicity by disintegrating when wetted,
thereby preventing birds from ingesting intact granules.

KEYWORDS: inert, granular, carrier, organic, agglomerated,
pesticides.

INTRODUCTION

Granular inert carriers used in the pesticide formulation industry can be divided into two broad categories based on their composition: inorganic and botanical. Ross (1) listed the more popular
carriers and their physical proerties under each of these two classifications.

Mr. Moechnig is a Research Engineer at Cargill, Inc., P. O.
Box 9300, Minneapolis, MN 55440; Mr. Wilhelm is Vice-President,
Fertilizer, Cargill, Inc., P. O. Box 9300, Minneapolis, MN 55440

The inorganic category is composed predominantly of the clays, with lesser amounts of sand and vermiculite. While low in cost, the clays often times need to be treated with a deactivator to prevent decomposition of the active ingredient prior to formulation. Van Valkenburg (2) has reviewed the deactivator requirements for different classes of chemicals.

The botanical inert carriers consist promarily of the corncob products, with lesser quantities of walnut shells, rice hulls, and wood used. Continued expansion of the use of combines to harvest corn, along with the drought of 1988, has raised concerns about the supply of corncobs and driven the price up.

The botanical carriers, along with the low-volatile matter clays, sand and vermiculite all break down slowly or not at all when applied to the soil. As a result, granules that remain intact and contain active ingredients can be available on the soil surface. Incorporation can significantly reduce, but does not eliminate, the presence of granules on the soil surface. These granules have reportedly been ingested by birds, causing their death and leading to action by EPA to ban or severely restrict the use of some products to mitigate this problem. Diazinon and Furadan are both currently under review by EPA as a result of the avian toxicity issue. An inert granular carrier that would maintain its integrity during formulation but disintegrate when wetted by rainfall or irrigation after application would be advantageous in reducing the likelihood of birds ingesting the granules.

OBJECTIVES

Due to the price and supply concerns over corncobs, Cargill's Fertilizer Division was approached by an outside company to provide a granular inert carrier with physical properties that would allow its use in place of corncobs. The following properties were desired in the final product.

Bulk density	0.48 - 0.56 g/cc
Absorptive capacity	25 - 40%
pH, 5% slurry	5 - 8
Resistance to attrition	>95%
Moisture content	<3%
Particle size	10/40 nominal

PROCESS DESCRIPTION

Based on laboratory and pilot plant trials, a carrier possessing the desired properties was developed. To produce the inert carrier, the starting material is first ground to pass a 35-mesh sieve. The ground material is then metered into an agglomeration unit where a suitable binder is applied to form green granules. These granules are then dried to the desired final moisture content and screened. Oversize product is reduced in size and routed back to the screen, while the undersized material is routed back to the agglomerator. The finished product is then conveyed to bulk storage bins.

PRODUCT CHARACTERISTICS

Laboratory trials were performed on samples of the raw material from different locations. Each of the samples was processed and the physical properties of each sample measured on the 10/40 mesh fraction of the resulting granular carrier.

BULK DENSITY

Free-fall bulk density was measured using ASTM Standard Test Method E727-80. Bulk density values ranged from 0.48 to 0.56 g/cc. Packed bulk density was measured by placing the graduated cylinder containing the carrier on a vibrating table for 60 seconds and reading the volume occupied. Bulk densities after packing ranged from 0.52 to 0.61 g/cc.

ABSORPTIVE CAPACITY

The absorptive capacity of the carrier was determined by weighing out 100 g of the granules and placing them in a pan. A small amount of water was misted onto the surface of the granules followed by gentle agitation to mix the water and carrier. This procedure was repeated until the granules began to stick together forming larger agglomerates. At this point, a sample was taken and the moisture content determined using an air-oven procedure. The moisture content, expressed as the weight of water removed in the drying process, divided by the dry weight of the granules, was taken to be the water absorptivity. Values of water absorptivity ranged from 29.9 to 47.5%.

BULK pH

The bulk pH of the carrier was measured using a pH meter on a slurry consisting of 5 g of carrier in 95 ml of deionized water. Measurements ranged from 4.4 to 4.7 on the samples produced.

RESISTANCE TO ATTRITION

The resistance to attrition was measured following ASTM Standard Test Method E728-80, using 13 mm steel balls. Values ranged from 95.0 to 99.0% in the samples tested. The 13 mm balls were used in place of the 16 mm balls called for in the standard due to the larger balls not being readily available at the time the tests were run.

WATER DISINTEGRABILITY

In attempting to measure the absorptive capacity of the carrier, it was found that the granules disintegrated when placed in water. When placed in a beaker of water and stirred, the granules break down within a few minutes to the original fine particles of the raw material used. As with the clay products, the rate and degree of disintegrability can be controlled by changes in the processing conditions, binders, and materials used. The carrier can be made to disintegrate rapidly (within a few minutes), or be essentially resistant to disintegration when placed in an aqueous media.

MOISTURE CONTENT

After agglomerating, the samples were dried in an oven overnight at 65 C, resulting in moisture contents below 3%. Control of the final moisture content in production to a desired level can be done by adjusting the drying air temperature and/or the residence time of the material in the dryer.

PARTICLE SIZE

Similar to the moisture content, particle size can be manipulated by changing process conditions to produce products in the mesh ranges typically specified for granular carriers. Adjusting the amount of binder used, making adjustments to the agglomerator, and post-drying screening and crushing can all be employed to arrive at the desired size range for the finished granule.

SUMMARY

An organic inert granular carrier has been developed with physical properties that make it appear to be a suitable replacement for corncobs or clays in the formulation of pesticide products. In addition to having good bulk density, water absorptivity, pH, and resistance to attrition, the new inert also disintegrates readily when placed in water. This property could have application in the formulation of pesticides that are applied to the soil surface by reducing the time that intact granules are available to be ingested by birds.

REFERENCES

(1) Ross, H. W., "Available Granular Carriers--Properties and General Processing Methods", Pesticide Formulations and Applications Systems: Second Conference, ASTM STP 795, K. G. Seymour, Ed., American Society for Testing and Materials, 1983, pp 32-44.
(2) Van Valkenburg, W., Pesticide Formulations, Marcel Dekker, Inc., New York, 1973.

Marvin L. Raymond

EVALUATION OF AIRBORNE DUST MEASUREMENT METHODS FOR AGRICULTURAL CHEMICAL CARRIERS

Reference: Raymond, M. L., "Evaluation of Airborne Dust Measurement Methods for Agricultural Chemical Carriers", Pesticide Formulations and Application Systems: 10th Volume, ASTM STP 1078. L. E. Bode, J. L. Hazen, and D. G. Chasin, Eds., American Society for Testing and Materials, Philadelphia, 1990.

ABSTRACT: Due to the environmental and safety concerns in the manufacturing and handling of toxic chemicals on inert carriers, the determination of the airborne dust from such materials is of interest. Therefore, a variety of airborne dust measurement methods have been studied to determine which ones are best suited for determining the dustiness of agricultural chemical carriers, and possibly serve as the basis for an ASTM standard procedure. The measurement techniques evaluated include light absorption, light scattering, gravimetric, and a charge transfer technique. The methods were evaluated for the following characteristics: ease of use, reliability, equipment required, sensitivity, and reproducibility. Descriptions of the equipment and techniques along with data are presented. Comments are made as to the appropriateness of the methods for determining airborne dust or comparing the dustiness of various carriers.

KEYWORDS: dust, airborne, carriers, laser, turbidity, triboelectric effect, gravimetric

Dust, the particles that become airborne when a granular material is poured or transferred in the open, is often taken for granted, but is becoming of greater concern due to possible adverse health effects. As previous authors have mentioned, dust in a carrier is usually considered as particles smaller than the limiting screen, which for most products would be minus 60 mesh or less than 250 micrometers [1,2].

Senior Research Scientist, Edward Lowe Industries, Inc. Research & Development, Cape Girardeau, MO 63701

However, the dust of concern is actually the smaller particles that remain airborne. Simply stated, dust is the cloud that remains airborne when a granular material is dropped from a given height. These particles are primarily less than 50 micrometers in diameter and many are less than 10.

Previous methods for determining the dustiness of a material have relied on dropping a relatively large sample into a chamber, such as described in ASTM Method for Index of Dustiness of Coal and Coke D547-41. This method involves a gravimetric determination of the dust collected after dropping 50 kg of the sample material through a chamber. Samples are rated according to an index of dustiness. The method of Goss and Reisch [2] as applied to carriers involves pouring a 4.5 kg sample into a chamber and detecting the dust with a tungsten light source. This method indicated good correlation between absorbance of light due to dust particles and the actual weight of the dust collected.

The purpose of this paper is to describe some methods which have been used to evaluate the dustiness of granular materials, particularly clay minerals, and report on more recent approaches that have been considered. The intent of these methods is to arrive at reasonably simple equipment setups that could easily be adapted to agricultural chemical carriers and used at a variety of laboratories.

DESCRIPTION OF METHODS

Laser Light Absorption Method

This method is based on the principle that a dust cloud created by dropping a sample from a certain height will block the transmission of a light beam such that the degree of dustiness can be determined from a light meter. This method is an adaptation of a previously published procedure referred to as an optical static dust tester (OSDT) [3]. The basic set up of the device is shown in Figure 1. The light source is a low powered laser that provides a coherent beam of light with a wavelength of 632.8 nm. A constant volume of granular sample is dropped from a height of about 46 cm above the laser beam. The beam is shielded on top such that only the dust cloud generated enters the beam. The instantaneous decrease of transmitted light due to any dust is displayed on the light meter and can be charted as a peak on a recorder.

This equipment can also be used in a different manner to monitor airborne dust that remains in the beam after the sample is dropped. In this procedure the shield is removed, the sample is dropped, and the time it takes for the dust to settle while the meter returns to 100% transmittance is measured.

Apparatus: The sample chamber consists of a Plexiglas cylinder of 10 cm in diameter and about 61 cm in height. The light source is a 1 milliwatt helium neon laser from Aerotech, Precision Laser Products, Pittsburg, PA 15238. The detector system consisting of a photodetector, meter, and power supply is a Sargent-Welch Chem Anal System. Part numbers are described in Figure 1. Signal output goes to a 100 millivolt strip chart recorder. The laser beam is positioned in the Plexiglas cylinder about 46 cm below the funnel from which the sample flows. The sample strikes a metal plate about 10 cm below the laser beam. The beam is shielded so that only the dust cloud that forms rises and enters the beam. The equipment should be mounted on a metal base or sturdy bench to prevent movement of parts during operation.

Procedure: A sample volume size of 150 ml (usually about 100 grams) is poured into the funnel at the top of the cylinder. A very low flow of air is purged around the windows near the laser beam source and the detector to prevent accumulation of dust. The instantaneous decrease of transmitted light due to any dust is displayed on the light meter and can be charted as a peak with the strip chart recorder.

A second technique that can be applied with this equipment is to measure the time required for the dust to completely settle under static air conditions. By dropping the sample directly through the laser beam, the airborne dust that remains in the light path can be detected. A recorder monitors the time required for the light meter to return to 100% transmittance. The time required is proportional to the amount of dust from the sample.

Light Scattering Method by Turbidity

This method is more of an indirect evaluation of the dust in that the airborne dust cloud that forms when a sample is poured into a chamber is swept with an air sampling pump into a collection vessel containing water. The water containing suspended dust particles is then evaluated in a turbidimeter [4]. The increase in turbidity is used as a measure of the dustiness of the sample. A diagram of the setup is shown in Figure 2.

Apparatus: The equipment required for this method consists of a Plexiglas cylinder (10 cm x 60 cm) with a funnel at the top and a hole in the side at about 20 cm in height from the bottom. The hole is large enough to accommodate a 37 mm filter cassette, such as Gelman #4338. A Staplex model TFIA high volume air sampling pump is connected via vacuum tubing (0.64 cm inside diameter) to a 500 ml filtering flask which is also connected to the filter cassette. The flask contains about 100 ml clean water. Turbidities are determined on a Hach turbidimeter, model 2100A.

Procedure: A 454 gram sample size is normally used to allow for a sufficient amount of dust to be collected in the water. However, smaller sample sizes down to 100 grams have been used. The sample is dropped from the funnel to the bottom of the cylinder in a continuous flow. As the sample hits a hard surface at the bottom of the cylinder, a dust cloud forms. The dust is continuously drawn into the water and the pump is left on for 30 seconds after dropping the sample. The flask containing the dust trapped in water is diluted to 300 ml with additional clean water. The water is then allowed to stand for 30 seconds to allow any larger particles (larger than 250 micrometers) that may have entered and are not considered airborne dust to settle out. A 25 ml sample is then taken and checked for turbidity.

Charge Transfer Method

This method is based on the principle that when solid materials rub together a charge transfer activity occurs. It is called the triboelectric effect since it involves friction. If an insulated probe is inserted into an air stream containing dust particles, the charge transfer that occurs when particles collide with the probe results in the generation of a minute triboelectric current. This current is directly proportional to the mass of the particles [5]. Although the equipment based on this principle was designed primarily for process streams to monitor dust, it was adapted to a laboratory setup for evaluation as an analytical tool.

Apparatus: The Triboflow Dust Monitor equipment consisting of a probe, venturi chamber to capture airborne dust, and associated electronics for measuring the degree of charge transfer was obtained from Auburn International, Danvers, MA   10923. The experimental setup is shown in Figure 3.

Procedure: Similar to the light absorption and light scattering techniques, a sample is dropped from a funnel into a Plexiglas chamber. With compressed air connected to the venturi chamber at a constant airflow of about 270 meters/min., the dust is drawn into the chamber past the probe. A meter on the control box shows the intensity of the signal based on the charge transfer effect. A totalizer counts from the time the dust first strikes the probe until the last dust particles pass through. A total count is obtained that is proportional to the concentration of the dust.

Gravimetric Method

One obvious way to quantitatively determine airborne dust is to actually collect the dust on a filter and weigh it. There are several ways in which this could be done, such as pouring a sample into a chamber while air sweeping the dust onto one or more filter cassettes. One unique method for which the equipment is commercially available involves rolling the sample in a chamber while drawing the dust out over a period of time. As the granular material is rolled, much of the granular surfaces are exposed to the air sweep allowing the dust to be removed and collected on a filter [6].

Apparatus: A device called a Dustmeter available from Heubach, Inc., Newark N.J. 07114 was used. This equipment was originally developed for determining the dustiness of powdered substances, particularly pigments. A diagram of the apparatus is shown in Figure 4.

Procedure: A 100 gram sample is placed in the metal chamber or dust generating container. This chamber rotates and has built in blades to simulate a conveying process on the sample material. The recommended velocity of air flow is 20 L/min with dust collected for a 5 minute period. Dust is collected on a 0.45 micrometer glass fiber filter paper (50 mm in diameter). The filter is weighed before and after dust collection to determine the amount of dust collected. This amount can be compared to the original sample weight to determine a percentage by weight.

Particle Size Analyzers

It has been suggested that commercial particle size analyzers on the market may offer a solution for evaluating airborne dust. Of course they are primarily designed for measuring the particle size distribution of very small particle size materials such as powders or microencapsulates. These have not been thoroughly evaluated for the purpose of this paper, but one device in particular was tested to determine how feasible this approach would be in discriminating between a dusty and non-dusty granular material.

The equipment evaluated was an Aerosizer Mach 2 manufactured by Amherst Process Instruments, Inc., Amherst, MA 01002 and available through Malvern Instruments, Southborough, MA 01772. This equipment is based on the measurement of the particles time of flight using light scattered from two laser beams. The dust particles are suspended in air and pass through a measurement zone. Particle size is determined from the acceleration rate. However, particles are also counted such that a total particle count is a function of the amount of dust generated from the granular material.

RESULTS

Various clay carriers were evaluated for dust by the different methods. The carriers differed in dustiness due to either the nature of the mineral or to dust suppression treatments given to some of the samples. A comparison of representative data collected is summarized in Table 1. In general, all the methods could distinguish between samples that had little airborne dust and those that were rather dusty. An organic fiber based granular carrier (25/50 mesh size) served as a reference for a sample with little or no dust. When the optical static dust tester (laser light absorption method) is used in the instantaneous mode, the cleaner samples appear as very similar. That is, it is difficult to distinguish degrees of dustiness of very clean samples. However, in the dust settling mode where more time is allowed in evaluating a sample, it becomes easier to distinguish some dust in the very clean samples. In either mode this method offers a good way to compare samples in terms of relative dustiness.

The gravimetric procedure offers a precise method for actually quantifying the amount of dust, but even with small samples (100 grams or less) it is not possible to actually collect all the airborne dust generated. It is also suspected that there is some attrition to the sample in the dust collection process, such that samples appear dustier in this method than in the other methods. This is shown by the unexpected high increase in sample weights for carriers G and H which were the more friable carriers.

The triboelectric method with the Triboflow dust monitor needs some additional work to control some of the variables, such as air flow. However, it is a rapid test like the laser method and can be made rather sensitive to dust. A very dusty sample or a larger sample size that generates a lot of dust can overload the probe or be out of range for a given sensitivity scale.

The turbidity method has been used extensively and makes a precise and reliable technique for checking samples routinely for relative dustiness. It correlates well with the optical static dust tester, though relatively clean samples can have turbidity readings of 1-10 units.

The Aerosizer particle size analyzer technique did differentiate between a clay carrier that had a dust suppression treatment and one that was not treated. The treated sample had a total dust particle count of 298,000 while the untreated (dusty) sample had a total count of 1,193,000 or 4 times greater. The dust particles observed were less than 5 micrometers. This device, however, has an equipment cost of about $35,000 compared to much lower costs for the other methods evaluated.

Table 2 summarizes attributes of the various methods.

CONCLUSIONS

A variety of methods have been evaluated for measuring airborne dust. At the very least, any of these methods are useful for comparing samples for dustiness relative to each other or to a standard material that does not generate dust. While a gravimetric procedure gives actual percentage dust by weight, the other methods should be considered further by those needing a rapid and relatively inexpensive means for determining the dustiness of granular carriers. The laser light absorption method (optical static dust tester) offers a reliable technique for monitoring airborne dust generated by a sample. The Triboflow equipment (triboelectric effect) also offers the potential for being a reliable technique, but needs additional research. For laboratories that already have a turbidimeter, this technique offers an easy way to monitor dust. It does involve more manipulations and thereby more room for errors.

It is also the intent of this paper to further the interest within ASTM of arriving at one or more standard methods for measuring airborne dust of agricultural chemical carriers. It is the author's opinion that the methods that would be best to pursue for standardization are the laser light absorption and the gravimetric methods.

ACKNOWLEDGMENTS

The author appreciates the assistance of Anne Spratt in testing some of the methods, Jim Job for the preparation of the illustrations, and Dana Bollinger for typing the manuscript.

REFERENCES

[1] Polan, J.A., "Formulation of Pesticidal Dusts, Wettable Powders and Granules," in Pesticide Formulations. W. Van Valkenburg, Ed., Marcel Dekker, New York, 1973, pp. 143-234.

[2] Goss, G.R. and Reisch, F.J. "A Technique for Dust Measurement," Pesticide Formulations and Application Systems: 8th Volume. ASTM STP 980. D.A. Hovde and G.B. Beestman, Eds., American Society for Testing and Materials, Philadelphia, 1988.

[3] Jelgic, M.F., "Laboratory Device for Evaluating Dust Control Agents", Journal of Coal Quality. July, 1986 pp. 104-107.

[4] "Turbidity Measurement" October 1985, Hach Company, P.O. Box 389, Loveland, CO 80539.

[5]   Dechene, Ronald L., "Triboelectric Flow Measurement",
      Measurements and Controls. February, 1988.
[6]   "Dustmeter - The Practical Method of Measurement to
      Ascertain the Dust Properties of Powderised and
      Granulated Substances", Heubach, Inc., Heubach Ave,
      Newark N.J. 07114.

TABLE 1 -- Method comparison data

Carrier Sample	Laser Light Absorption			Turbidity	Charge Transfer	Gravimetric
	%T	Abs	Dust settling time (sec)	Units	Counts	wt. %
A. 25/50 mesh organic fiber base	100	0	60	0	0	0.015
B. 16/30 mesh clay treated for dust	99	0.004	50	5	0	0.020
C. 16/30 mesh clay low dust	99	0.004	...	9	0	0.019
D. 25/50 mesh clay low dust	96	0.018	200	14	2	0.017
E. 25/50 mesh LVM clay	95	0.022	320	12	10	0.025
F. 16/30 mesh LVM clay	90	0.046	...	18	14	0.027
G. 16/30 mesh clay-high attrition	86	0.066	...	24	16	0.065
H. 25/50 mesh clay-high attrition	83	0.081	370	25	20	0.065

TABLE 2 -- Method attribute comparisons

Method	Ease of Use	Time Per Sample	Reproducibility	Sensitivity	Overall Reliability	Approximate Equipment Cost
Laser Light Absorption (OSIM)	Very easy. Does require 30 minute warm up time for stability. Need stable bench for support.	less than 5 min. 5-10 min. for monitoring dust settling time.	Relative Standard Deviation (RSD) normally within 10%.	Very sensitive to dust. Discriminates well between samples as to relative dustiness.	Reliable indicator of dust regardless of sample size or color. Some fluctuations due to instability.	less than $4000.00
Turbidity	Several steps involved, but fairly easy. Requires standardization of meter.	10 minutes	RSD within 10%	Sensitive, but dependent on collecting representative dust sample by air sweep.	Appears to be a good routine method. Sample color or mineral carriers containing high amounts of mica can be a problem.	less than $2500.00
Charge Transfer	Very easy. Need to adjust sensitivity on meter to keep samples in range.	less than 5 min.	RSD 10-20%	Very sensitive Loses sensitivity with very dusty samples.	Needs additional research. Airflow variation can be a problem.	less than $3000.00
Gravimetric (Dustmeter)	Easy to use.	10 minutes	RSD normally within 5%	Discriminates between samples, and appears to collect a representative sample of actual airborne dust.	Equipment reliable, but rolling technique may be breaking some samples down due to attrition and giving inaccurately high results.	$8000.00

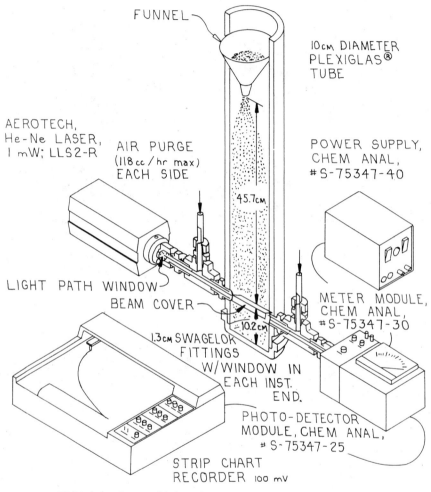

FUNNEL

10cm DIAMETER
PLEXIGLAS®
TUBE

AEROTECH,
He-Ne LASER,
1 mW; LLS2-R

AIR PURGE
(118 cc / hr max)
EACH SIDE

POWER SUPPLY,
CHEM ANAL,
S-75347-40

45.7cm

LIGHT PATH WINDOW
BEAM COVER

METER MODULE,
CHEM ANAL,
S-75347-30

10.2cm

1.3cm SWAGELOK
FITTINGS
W/WINDOW IN
EACH INST.
END.

PHOTO-DETECTOR
MODULE, CHEM ANAL,
S-75347-25

STRIP CHART
RECORDER 100 mV

FIG. 1 -- Laser light absorption method apparatus

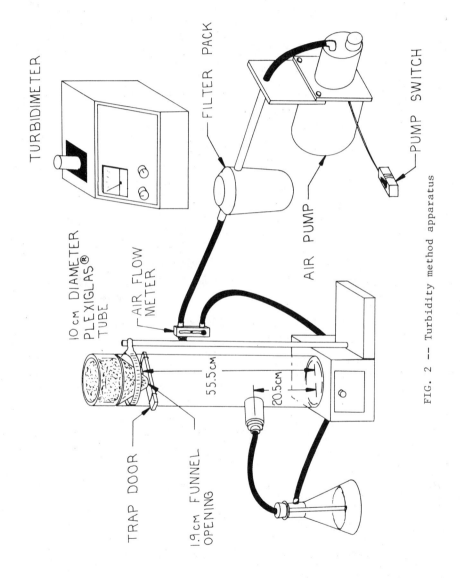

TURBIDIMETER

FILTER PACK

PUMP SWITCH

AIR PUMP

10 cm DIAMETER PLEXIGLAS® TUBE

AIR FLOW METER

55.5 CM

20.5 CM

TRAP DOOR

1.9 cm FUNNEL OPENING

FIG. 2 -- Turbidity method apparatus

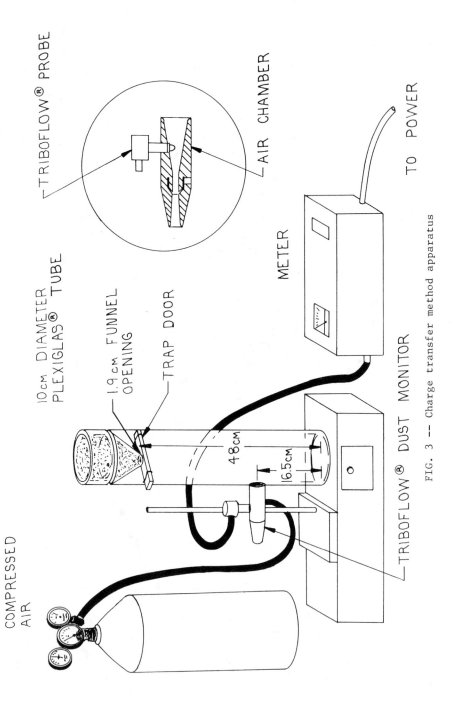

FIG. 3 -- Charge transfer method apparatus

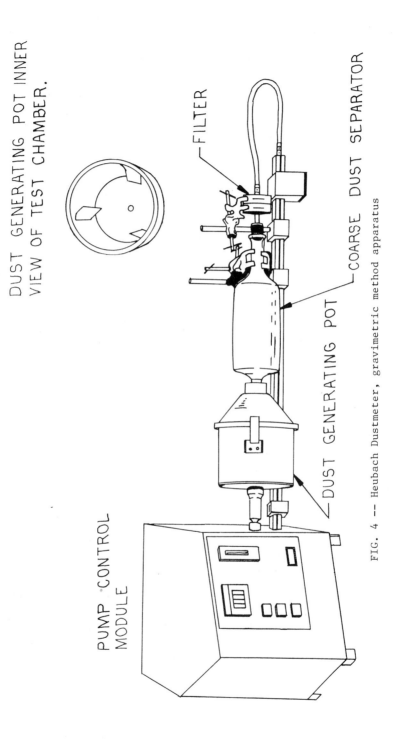

DUST GENERATING POT INNER
VIEW OF TEST CHAMBER.

FILTER

COARSE DUST SEPARATOR

DUST GENERATING POT

PUMP CONTROL
MODULE

FIG. 4 -- Heubach Dustmeter, gravimetric method apparatus

Randall G. Luttrell and David B. Smith

EFFECT OF SPRAY DEPOSIT CHARACTERISTICS ON INSECTICIDE EFFICACY

REFERENCE: Luttrell, Randall G., and Smith, David B.,
"Effect of Spray Deposit Characteristics on Insecticide
Efficacy," Pesticide Formulations and Application Systems:
10th Volume, ASTM STP 1078, L. E. Bode, J. L. Hazen, and D.
G. Chasin, Eds., American Society for Testing and Materials,
Philadelphia, 1990.

ABSTRACT: Dosage, droplet size, deposit density, and tank-
mix concentration are spray deposit characteristics which
may influence the efficacy of insecticides. The relative
influence of each of these variables on insect control is
difficult to measure because of the interrelationships among
the variables. Only a few experiments have actually
attempted to quantify the relative influence of all of the
spray deposit variables on insecticide efficacy. An
indirect approach to estimating the relative influence of
these variables on efficacy of insecticides for control of
Heliothis spp. larvae on cotton was developed and tested
over the past 5 years at Mississippi State University.
Results suggest that dosage is more important than the other
deposit variables in influencing Heliothis spp. control on
cotton. Implications of these results and limitations to
current research approaches are examined in this paper.

KEYWORDS: spray deposit characteristics, insecticide efficacy, dosage, droplet size, deposit density, tank-mix concentration

Application techniques potentially affect insecticide efficacy
either by influencing the amount of insecticide deposited to the
target area (deposition efficiency) or by altering the distribution
and concentration of the insecticide on the target area (deposit
characteristics). For a given pest-crop-pesticide-carrier, the spray
deposit characteristics which potentially affect efficacy are dosage
(amount of insecticide per unit area), droplet size, deposit density
(number of droplets per unit area), and tank-mix concentration (amount
of insecticide per unit volume of spray deposit). Both deposition
efficiency and the characteristics of the deposit on the target area
are potentially influenced by a wide range of factors including operational (atomizer type, flow rate, atomizer spacing, boom height,

Drs. Luttrell and Smith are Associate Professor of Entomology and
Professor of Agricultural and Biological Engineering, respectively;
Mississippi State University, Drawer EM, Mississippi State, MS 39762.

atomizer pressure, and ground speed), formulation, and meteorological variables. Although factors influencing deposition efficiency also affect deposit characteristics, a quantitative understanding of the importance of deposit characteristics on insect control would help define the acceptable range of spray deposit variables important to insect control. For example, if a wide range of droplet sizes provided equal levels of insect control, researchers could concentrate on designing application systems which produced droplet sizes optimum for increased deposition efficiency. However, if insect control is greatly influenced by a rather small range of droplet sizes, researchers will have less latitude in designing application equipment for optimum deposition efficiency. This same logic can be applied to all of the deposit characteristics.

With this rationale, a cooperative research program was initiated to define the relative importance of the various spray deposit characteristics on control of Heliothis spp. on cotton. An indirect, laboratory-bioassay approach for estimating the relative influence of deposit characteristics on insect control was developed (1, 2). Efforts to validate results of this indirect approach were made during the past 3 growing seasons in field experiments. In this paper we examine the advantages and limitations of different research approaches to studying effects of deposit characteristics on insecticide efficacy, report results of our studies, and consider potential implications of current results on future research.

MEASURING THE RELATIVE INFLUENCE OF DEPOSIT VARIABLES

Many research articles cite deposit characteristics as factors influencing the efficacy of insecticide treatments, but few actually measure the relative influence of the various deposit variables. Several studies (3, 4, 5, 6, 7, 8, 9, 10, 11) have examined the effects of one, two, or three of the four deposit variables on insect control, but we have found only six experiments (1, 2, 12) which include all four of the deposit variables in the experimental design. Most insect control studies are evaluated on the basis of insecticide application rate (IAR) with no measurement of insecticide deposition rate (IDR) to the target area. This is largely due to experimental problems associated with measuring the relative influence of variables which are interrelated.

Measurements of the relative effects of the different deposit variables on efficacy of insecticides requires carefully designed and vigorously controlled experiments. Direct approaches, such as studies conducted in field environments, are often confounded with several uncontrolled variables which typically prevent adequate measurement of treatment effects. Indirect approaches, such as the measurement of single or multiple variable effects and development of predictive relationships, eventually require validation in the highly variable field environment. Thus, accurate estimates of the relative influence of deposit characteristics on efficacy of insecticides are difficult to obtain.

Confounded Nature of Deposit Variables

Probably the most limiting factor to measuring the relative influence of deposit variables on insecticide efficacy is the interrelationships among the variables. For a given volumetric application rate, droplet size will increase as deposit density decreases. Droplet size and deposit density account for the volumetric deposition rate delivered to a given target area. Dosage on the target area is directly influenced by the concentration of insecticide in the tank mixture and the total volume of spray deposited. Thus, changes in any one of the four deposit characteristics will alter one or more other deposit characteristics. This creates tremendous experimental problems both from the perspective of manipulating a range of treatments and in data analyses.

We used a multiple regression procedure to study the relative importance of interrelated variables (1, 2). Because of the confounded nature of the variables in these experiments (1, 2), we have been extremely careful to report the interrelationships between the various deposit variables and include results of correlation analyses in our publications. For example, combinations of dependent and independent variables were significantly correlated with each other in the data set used to develop a regression model for fluvalinate-water sprays (2). Data interpretation must be approached carefully because of these interrelationships. Relative comparisons in these studies were made on the basis of standardized regression coefficients which describe the amount of total variation explained by each variable in the resulting regression models. Since these coefficients are not direct comparisons between or among the variables, the importance of each variable was related to the most important variable in each data set. Dosage has consistently been the variable which explains the largest amount of variation in insect mortality in our previous studies.

Measurement of Physical Characteristics and Biological Influences

In addition to the experimental problems associated with estimating the relative influence of interrelated variables on insecticide efficacy, accurate measurements of the four deposit variables are difficult, time-consuming, and expensive. In attempts to measure deposit characteristics in field studies, we found the field environment to be influenced by external sources of variation large enough to prohibit delivery of our experiment treatments to the target area (i.e. we were not able to control volumetric deposition rates with sufficient confidence to experimentally manipulate treatments). To gain more precise control of volumetric deposition rate, we applied our treatments to soybean leaflets collected from field plots and pinned to a spray stand. Applications were made within an enclosed building to reduce environmental influences on volumetric deposition rates. Because regression techniques were used to study the interrelationships among the various deposit variables, it was essential that a wide range of volumetric deposition rates be obtained. Such large data sets require experimental control of treatment effects.

Problems were also experienced in measuring dosage, droplet size, and deposit density directly on the plant foliage. Analytical

techniques, such as gas chromatography, are probably the ideal methods
for measuring dosage, but the large number of samples necessary to
develop regression models were cost prohibitive. It is also difficult
to directly measure droplet size and deposit density on cotton
foliage. Several previous researchers (6, 11, 13, 14) have recognized
this problem. Himel (14) developed a fluorescent particle method to
study spray deposits. Because of potential adhesion problems with the
fluorescent particle method, we chose to use a fluoremetric technique
described by Smith et al. (15). Initial efforts to obtain dependable
fluorometric readings from samples extracted from cotton leaf surfaces
were variable, perhaps because of the various pigments contained in
cotton leaves. We have recently found that careful selection of
fluorescent dyes for a given light spectrum may over-come this
problem. Because of the problems associated with measurements of
deposit density and dosage on leaf surfaces, an artificial target was
used to measure deposit characteristics.

Insect mortality was measured by placing ten neonate Heliothis
virescens larvae on leaf disks (11.4 cm^2) taken from soybean leaflets.
A glass cover-slip, the artificial target, was placed adjacent to the
soybean leaflet to collect spray deposits for dosage measurements. The
density of insects used was much higher than that which would occur in
field situations, but necessary because of the need to have a measurement of insect mortality for each corresponding measurement of dosage,
droplet size, and deposit density. Limitations to this technique seem
to be closely associated with the behavior of the insect at rather
high densities within a relatively small area. We assumed that this
technique would mimic the behavioral relationships of these larvae as
related to the range of deposit characteristics tested. Previous studies (4, 5, 6) suggest that Heliothis virescens larvae can detect
large droplets (> 700 um in diameter) and avoid them. While we
recognize these reports, the droplet sizes used in these studies were
much larger than those typically encountered in production agriculture
or used in our studies. However, insect behavior as related to deposits of insecticide is poorly understood. It seems reasonable to
expect different behaviors with different insect-crop-insecticide
relationships.

Definition of Spray Target

Accurate definition of the spray target is an essential component
in pesticide application research. Once an insecticide is applied to
a crop, the probability of an insect contacting a toxic dosage is
influenced by insect behavior (feeding, locomotion, and sensory),
toxicity of the particular insecticide to the particular population of
insects present, mechanism of insecticide activity (contact, consump-
tion, or fumigant), plant growth, pesticide characteristics which
alter the relative concentration of the the deposit through time, and
environmental factors (ultra-violet light, temperature, rain, wind,
etc.) which alter the persistence of insecticide sprays.

A limited number of studies have been conducted to measure the
relative importance of different mechanisms of contact with insecticide by cotton insects. Scott et al. (16) studied the effects of
both direct impingement (spray directly contacts pest during application) and residual contact (pest encounters pesticide deposit on the

plant after application) mechanisms for control of boll weevil
(Anthonomous grandis Boheman) with azinphosmethyl. Their data
suggested that 80% of boll weevil control is due to weevil contact
with azinphosmethyl via the residual contact mechanism. Wofford (17)
studied the direct impingement and contact mechanisms of insecticide
mortality of Heliothis virescens larvae exposed to permethrin on cotton terminal buds. Larvae were placed on sprayed terminal buds either
before (direct impingement + residual contact) or after (residual contact) the application process. In these studies, 84% of the total
mortality could be explained by residual contact. MacQuillan et al.
(18) reported similar results in studies using Heliothis punctigera
and sprayed tobacco leaf discs. Based on ratios of their dosage mortality studies, one can assume that 78% of the total larval mortality
observed was due to residual contact.

We have conducted studies similar to those reported by Wofford
(17) with different instars of Heliothis virescens larvae exposed to
cotton terminal buds treated with permethrin--water sprays (Table 1).

TABLE 1 -- Mortality of Heliothis virescens larvae exposed
to cotton terminal buds sprayed with permethrin before
(impingement + residual mechanism of contact) and after
(residual contact) the application.[a]

Mean Percent Mortality[b]

Mechanism of Contact[c]

Larval Instar	Impingement + Residual	Residual	Percent of Total Mortality Due to Residual Mortality[d]	t values[e]
1	100.0	90.5	90.5	9.92**
2	83.5	65.3	78.1	6.18**
3	72.0	45.5	63.6	5.48*
4	62.8	34.5	56.6	4.23**
5	54.0	23.3	45.9	5.84**

[a]Permethrin was applied in a water spray at a volumetric application
rate of 46.8 1/ha and a toxicant application rate of 0.03 kg active
ingredient/ha.
[b]Mortality was determined at 72 h posttreatment and was corrected for
mortality in the untreated check.
[c]Larvae were placed on plants 30 minutes before (impingement + residual) and 30 minutes after (residual) application.
[d]Determined by (% residual mortality / % impingement + residual mortality) * 100.
[e]* and ** refer to P < 0.01 and P < 0.001, respectively.

Results were essentially the same as those reported by Wofford (17) in
that residual contact appears to be the most important mechanism of
insecticide contact for Heliothis virescens larvae on cotton.
Significant differences were detected between impingement + residual

and residual contact mechanisms for all instars. Residual contact
became less important as larval instar (size) increased. This may be
influenced by our bioassay procedures which more closely relate to
normal field environments for smaller larvae. Large larvae are typically located in fruiting structures lower in the plant canopy.
Significant direct impingement is unlikely at these locations because
only a small amount of the total spray would reach these locations
during typical field applications.

Collectively these data suggest that residual mechanisms of contact are more important than impingement mechanisms for control of
Heliothis virescens larvae on cotton with contact insecticides such as
permethrin. Other pest-crop-insecticide situations would not
necessarily follow the same pattern. However, these studies support
the use of an indirect approach to measuring mortality (i.e. bioassays
on treated plant tissue) which utilizes residual mechanisms of contact
with the insecticide. The general lack of information on insect behavior and relationships with insecticide contact indicate that additional research is needed with many insect-crop-insecticide
situations. Utilization of indirect approaches to studying deposit
characteristics assumes some understanding of the spray target.
Detailed definition of optimum spray targets is undoubtedly a key
element for future refinement of application techniques.

Direct Versus Indirect Approaches

Because of our inability to control all essential experimental
variables in a field environment, we used indirect approaches to
study the effects of deposit characteristics on insect control. These
indirect approaches provided us with the rather elaborate data sets
necessary to partition the variation in mortality associated with each
independent variable. However, the numerous limitations to this
research approach, as described above, emphasize the need to validate
results in field studies.

Obtaining reliable estimates of absolute population densities of
insects at low densities normally encountered in field crops is difficult. Cotton is routinely monitored for the presence of Heliothis
spp. and control recommendations are usually based on sample estimates. However, these estimates of actual population densities are
generally extremely variable. It is not unusual for the sample data
to have standard deviations greater than the mean. While these
sampling techniques seem to be too variable for some experimental purposes, they are reliable enough to time applications of highly effective insecticides. The application literature contains many examples
of the excessive rates of insecticide used in production agriculture.
Generally, researchers cite these excessive use rates as reasons for
improved application procedures. It also appears that excessive use
rates of insecticide may mask poor management decisions based on
extremely variable sample information. The lack of accuracy in estimating absolute population densities limits research designed to
measure treatments which have small effects on insect mortality.

Differences of 20% mortality in Heliothis virescens larvae are
extremely difficult to measure in field situations. Because of the
amount of economic damage caused by a single Heliothis spp. larva,

insecticide applications are recommended at rather low population densities. The recommended treatment threshold for control of Heliothis spp. larvae on cotton is when 4% of the plants are infested (19). At a normal plant density of 100,000 plants/ha, the economic threshold is 4,000 larvae/ha. Confidence limits (95%) for much of our field data included insect densities twice the average density of ca. 2,000 larvae/ha. Thus, for an average insect density of 4,000 larvae/ha we could expect to detect significant differences when treatments reduced the insect density to ca. 0-800 larvae/ha. This would require mortality rates of 80-100%. Thus, the probability of detecting differences in treatments causing less than 20% differences in mortality are unlikely.

Another experimental problem in validating bioassay results is the ability to actually deliver needed insecticide deposition rates to the target area. While one can easily calculate and mix a range of insecticide application rates for a given set of application variables, physical characteristics of the spray may affect the deposition efficiency of the application. With our indirect approach we were able to maximize deposition efficiency in the enclosed environment. Manipulating insecticide deposition rates in the field environment is extremely difficult and elaborate controls on treatment effects are necessary. Because of expected differences in deposition efficiencies for various droplet sizes, field validation studies would require careful monitoring of dosage, droplet size, and deposit density. Problems associated with monitoring these variables in field environments were a primary influence on the initial development of our indirect research approach.

These comments are not intended to discourage field studies. Field validation of results obtained by indirect approaches is essential if the results are to be transferred to production agriculture. Indirect approaches, such as our bioassay procedure, can often be used to measure the relative significance of experimental variables and determine the need for field testing. In our studies, deposit characteristics other than dosage appear to have small effects on insect control and detection of the effects of other deposit variables on insect control in field environments would not be expected. These conclusions should not be expanded to other insect-crop-insecticide situations or to values for each deposit variable outside the range of those included in our studies.

RELATIVE IMPORTANCE OF SPRAY DEPOSIT CHARACTERISTICS

For several years, we have conducted research to understand the relative importance of deposit characteristics on cotton insect control. Since few studies had previously attempted to define the relative influences of dosage, droplet size, deposit density, and tank-mix concentration on insect mortality, research techniques which would allow us to accurately measure or estimate all four of the deposit variables simultaneously and relate these measurements to insect control were not readily available. Because of the interrelationships of the variables to be studied and problems associated with direct measurements in field environments, an indirect research approach was used to estimate effects of deposit variables on insect mortality.

Regression techniques were used to describe data obtained from these insect bioassay experiments where dosage, droplet size, deposit density, and tank-mix concentration were observed for a range of insect mortalities. Detailed descriptions of these procedures were made by Wofford et al. (1) and Smith and Luttrell (2). The ranges of values for each variable studied are listed in Table 2.

TABLE 2 -- Ranges of independent variables used to develop regression models of larval mortality as a function of dosage (DO), droplet size (DS), deposit density (DD), and tank-mix concentration (TM).[a]

Insecticide--Carrier	DO	DS	DD	TM
permethrin--water	0.1-213	89-336	0.80-387	0.9-15.4
permethrin--soybean oil	0.1-26	24-413	0.10-198	5.8-23.2
fluvalinate--water	0.007-74	97-390	0.08-201	0.25-1.5
fluvalinate--cottonseed oil	0.1-213	42-299	0.33-321	4.5-22.7

[a] Dosage (DO) expressed in g active ingredient/ha; droplet size (DS) in volume mean diameter (um); deposit density (DD) in drops/cm2; and tank-mix concentration (TM) in g active ingredient/l of finished spray.

Using the indirect bioassay approach, regression models for four separate insecticide-carrier combinations; permethrin in soybean oil, permethrin in water, fluvalinate in cottonseed oil, and fluvalinate in water (Table 3); were constructed. Mortality data of 100% were eliminated from the analyses, and mortalities of 0% were reassigned as 1% mortality for development of the regression models.

By comparing standardized regression coefficients, the relative influence of each deposit variable on insect mortality (Table 4) can be estimated. In all four models, dosage was the most important variable influencing mortality. In one model, permethrin-soybean oil, droplet size was nearly as important as dosage. When the relative influence of each deposit variable is examined across all four data sets, dosage appears to be much more important than the other three deposit variables. However, the relative importance of spray deposit characteristics may be a function of the carrier and particular pesticide used. If so, these results may differ from those for other carrier-pesticide combinations. It is significant to note that all four deposit variables had significant influences on insect mortality in one or more of the insecticide-carrier experiments.

Field studies were conducted to validate the effects of dosage on cotton insect control in 1987 and 1988. Since dosage was a major component of all regression models and the magnitude of the effect was relatively large, chances of measuring treatment influences in the field environment were higher than those expected from other deposit variables. In these studies several insecticide application rates of

TABLE 3 -- Regression models developed to describe mortality of
Heliothis virescens larvae as a function of dosage (DO),
droplet size (DS), deposit density (DD), and
tank-mix concentration (TM).[a, b, c]

---

permethrin -- water

$$n = 147 \qquad R^2 = 0.6403** \qquad F = 84.8***$$
$$k=0.89055-0.15864(\log DO)2+0.23155(\log DD)-0.01980(TM)$$
$$MO = (10^k*100)/(1+10^k)$$

permethrin -- soybean oil

$$n = 222 \qquad R^2 = 0.4963** \qquad F = 71.6***$$
$$k = 3.1856+1.00655(\log DO)-0.17965(\log DD)-0.000012403(DS)^2$$
$$MO = (10^k*100)/(1+10^k)$$

fluvalinate -- water

$$n = 223 \qquad R^2 = 0.473** \qquad F = 198.8***$$
$$k = -1.855+1.115(DO)^{0.3}$$
$$MO = (10^k*100)/(1+10^k)$$

fluvalinate -- cottonseed oil

$$n = 241 \qquad R^2 = 0.414** \qquad F = 84.0***$$
$$k= -0.374+0.440(\log DO)+0.547(\log TM)$$
$$MO = (10^k*100)/(1+10^k)$$

---

[a]Taken from Wofford et al. (1) and Smith and Luttrell (2). Regression
equations were developed by using the multiple linear regression
method (20). Mortality data were studied in the form of logit mortality (21).

[b]Log implies logarithm to the base 10.

[c]** and *** refer to P < 0.01 and P < 0.001, respectively.

fluvalinate were applied to small plots (4 rows by 15 meters) of cotton. The application rates and dates varied in 1987 and 1988. In
1987, fluvalinate was applied at rates of 11, 22, and 44 gm/ha on July
10, July 15, July 20, July 27, and August 3. Based on our regression
models, these rates should have provided insect mortality rates of 44,
73, and 92 percent, respectively. In 1988, fluvalinate rates of 2, 6,
10, 16, 25, 40, and 148 gm/ha were applied to a field of late-maturing
cotton on August 17, August 23, August 30, and September 7. Predicted
mortalities were 15, 28, 42, 56, 70, 84 and 99 percent, respectively.
All treatments were replicated 8 times in 1987 and 12 times in 1988.
Plots were monitored weekly by standard field scouting procedures.
Data were collected on insect density and plant damage found in the
terminal bud, in squares, and in bolls. In each plot, 25 randomly
chosen plants were searched.

Dosage was correlated with most of our measurements of insect mortality (Table 5). While we were able to correlate insect density and
plant damage with predicted mortality, significant differences between

TABLE 4 -- Influence of droplet size, deposit density,
and tank-mix concentration relative to dosage on
mortality of Heliothis spp. larvae.

Relative Influence on Heliothis spp. Mortality[a]

Insecticide-Carrier	DO	DS	DD	TM
Permethrin-Soybean Oil	1.00	0.94	0.31	0.003
Permethrin-Water	1.00	0.00[b]	0.28	0.16
Fluvalinate-Cottonseed Oil	1.00	0.00[b]	0.14	0.55
Fluvalinate-Water	1.00	0.07	0.26	0.16
Average	1.00	0.25	0.25	0.22

[a]Dosage (DO) expressed in g active ingredient/ha; droplet size (DS)
in volume mean diameter (um); deposit density (DD) in drops/cm2; and
tank-mix concentration (TM) in g active ingredient/l of finished
spray. Estimated values relative to dosage. Based on standardized
regression coefficients reported by Wofford et al. (1) and Smith and
Luttrell (2).
[b]Listed as 0.00 because the variable would not enter the regression
equation due to small tolerance.

treatments at 1X and 2X rates were not detected. This illustrates the
problem of measuring effects of treatments causing relatively small,
differential amounts of insect mortality in field experiments.
Observed variables that were more closely associated with insect
mortality on the date of application (i.e. number of Heliothis larvae
in terminals and squares) or only a few days after application
(damaged terminals, damaged squares) tended to be more closely
correlated with predicted mortality than those variables which were
potentially influenced by other factors (damaged bolls, yield).
Logically, one would expect more external variation (i.e. plant damage
from other insects, environmental and cultural effects on plant fruit
production) when the time between cause (mortality of larvae at
application) and effect (yield) was increased. Measurements taken
closer to application (damaged squares) should correlate better with
predicted mortality.

Since all of the regression models developed by the indirect
approach indicated that dosage was the most important variable (Table
4) and dosage effects correlated well with variables used to estimate
efficacy of insecticide treatments in field studies (Table 5), we
concluded that dosage is the deposit characteristic of most importance for the pest-crop-insecticide combination we studied. Further
examination of the regression models illustrates the magnitude of the
influence that deposit variables other than dosage have on insect
mortality. Figure 1 was prepared by plotting the predicted
mortalities obtained from the regression models developed and reported
by Wofford et al. (1). To illustrate the maximum effect of deposit
variables other than dosage on insect mortality, predicted mortalities
were obtained for best and worst case scenarios within the range of
data included in studies by Wofford et al. (1) (Table 2) and across a

Table 5 -- Relationship between predicted mortality of larvae
and observed insect and plant damage parameters measured in
small-plot, field studies on cotton.[a]

| | Correlation With Predicted Mortality | | | |
| | 1987 Studies | | 1988 Studies | |
Variable Measured[b]	n	Correlation Coefficient[c]	n	Correlation Coefficient
Heliothis eggs in terminal buds	40	0.201	96	0.365**
Heliothis larvae in terminal buds	40	-0.675***	96	-0.278*
Damaged terminal buds	40	-0.780***	96	-0.176
Heliothis damaged buds	40	-0.520***	96	-0.601***
Weevil damaged squares	40	-0.359*	96	0.046
Heliothis larvae in square	40	-0.525***	96	-0.517***
Heliothis damaged bolls	40	-0.264	96	-0.222*
Heliothis larvae in bolls	40	-0.045	96	0.503***
Yield	40	0.090	96	0.088

[a]Plots were treated with different rates of fluvalinate to give predicted mortality rates of 0, 15, 28, 42, 56, 70, 84, and 99%. The
experimental design included 8 replications of each treatment in
1987 and 12 replications of each treatment in 1988.
[b]Seasonal averages were used in the correlation analyses. All data
were recorded and analyzed on the basis of the number observed per
25 plants.
[c]*,**, and *** refer to significant (P 0.05), highly significant
(P 0.001), and very highly significant (P 0.0001) correlations,
respectively.

wide range of dosages. For the permethrin-soybean oil model, droplet
size was the second most important variable. The remaining variables
in each model were adjusted to conform to typical field application
characteristics since changes in a particular independent variable
necessitates changes in another independent variable. For example,
changes in droplet size require changes in deposit density at the same
volumetric application rate (VAR). All simulations were made assuming
a deposition efficiency of 85%. Results of these comparisons indicate
that at dosages approaching recommended insect control rates (0.1
kg/ha) difference in predicated mortality are small regardless of the
application procedure. At lower dosages and predicted mortalities,
the differences between best and worst case scenarios become more
evident.

However, the maximum amount of difference at a given dosage is
less than 20% mortality for a given insecticide-carrier. Based on
these relatively small amounts of difference in predicted mortalities
and the need to illustrate these differences at dosages much less than
those typically applied in production agriculture, we conclude that
further refinement and measurement of the impact of deposit variables

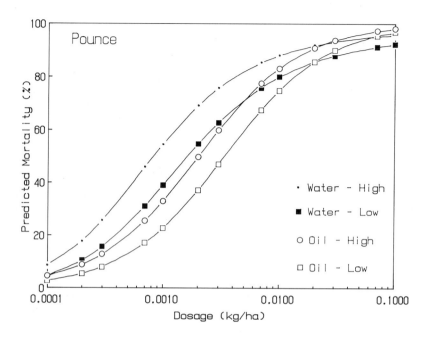

Fig. 1. Dosage–mortality, relationships predicted from
regression models constructed to describe the effects
of spray deposit characteristics on mortality of
Heliothis virescens larvae exposed to cotton
with pyrethroid insecticides.

on insecticide efficacy seems unnecessary. This conclusion is limited
to the particular insect-crop-insecticide combinations studied and the
ranges of values used for each deposit variable studied (Table 2).
Insect mortality, the dependent variable, ranged from below 10% to
above 90% in all four data sets.

SUMMARY AND CONCLUSIONS

Based on the results obtained in our laboratory and field studies
during the past 5 years, we found that dosage is the most important
variable influencing control of Heliothis spp. larvae on cotton with
contact insecticides. The range of values for the four deposit
variables were within the range of currently used insecticide applications in production agriculture. It is extremely important to
remember that the effects of dosage, droplet size, deposit density,
and tank-mix concentration are confounded and difficult to separate.
Results obtained with Heliothis virescens larvae, cotton- and soybean-
leaflets, and contact insecticides would not logically expand to other
insect-crop-insecticide situations. As indicated throughout the text,
insect behavior has a major influence on contact with insecticides.
Our studies were conducted with a mobile insect. Results may differ
from systems associated with insects exhibiting different behaviors.

For example, aphids tend to remain at the same general location on host plants. The probability of an aphid contacting spray deposits of contact insecticides would be drastically different from a Heliothis virescens larva. Hall and Reichard (20) have studied the effects of deposit characteristics on mortality of two-spotted spider mites.

Given the major influence of dosage in our studies, we believe that research should concentrate on optimizing deposition efficiency to the target area. Additional research should be conducted on defining optimum spray targets and understanding relationships between insect behavior and probability of insecticide contact. For Heliothis spp. control on cotton, our studies suggest that future experiments should include measurements of dosage and corresponding levels of insect control in their experimental design. Because of the lack of literature which includes accurate estimates of dosage effects on insect control and the major role of dosage exhibited in our studies, we recommend that all future studies designed to study effects of deposit variables on insect control include accurate measurements of dosage. Simple estimates of dosage based on insecticide application rates seem to be inappropriate for these types of studies.

The complex interrelationships between biological and physical variables in insecticide application research creates a need for diverse scientific expertise and cooperation among different academic disciplines. We believe that our interdisciplinary approach has been fruitful, and we strongly encourage cooperative research between the biological and physical sciences to further expand application research.

REFERENCES

(1) Wofford, J. T., Luttrell, R. G., and Smith, D. B., "Relative Effect of Dosage, Droplet Size, Droplet Density and Droplet Concentration on Mortality of Heliothis virescens (Lepidoptera: Noctuidae) Larvae Treated with Vegetable Oil and Water Sprays Containing Permethrin," Journal of Economic Entomology, Vol. 80, 1987, pp. 460-464.

(2) Smith, D. B. and Luttrell, R. G., "Performance Specifications for Tobacco Budworm (Lepidoptera: Noctuidae) Larvae Treated with Vegetable Oil and Water Sprays Containing Fluvalinate," Journal of Economic Entomology, Vol. 80, 1987, pp. 1314-1318.

(3) Awad, T. M. and Vinson, S. B., "The Pickup and Penetration of Ultra-Low-Volume and Emulsifiable Concentrate Malathion Formulations by Tobacco Budworm Larvae," Journal of Economic Entomology, Vol. 61, 1968, pp. 242-245.

(4) Polles, S. G., "The Effect of Various Droplet Sizes of ULV Insecticide Concentrate on Residual Action and Toxicity," M.S. Thesis, Mississippi State University, Mississippi State, 1968.

(5) Himel, C. M., "The Optimum Size for Insecticide Spray Droplets," Journal of Economic Entomology, Vol. 62, 1969, pp. 919-925.

(6) Burt, E. C., Lloyd, E. P., Smith, D. B., Scott, W. P., McCoy, J. R., and Tingle, F. C., "Boll Weevil Control with Insecticide Applied in Sprays with Narrow-Spectrum Droplet Sizes," Journal of Economic Entomology, Vol. 67, 1970, pp. 365-370.

(7) Wolfenbarger, D. A., and McGarr, R. L., "Low-Volume and Ultra-Low-Volume Sprays of Malathion and Methyl Parathion for Control of Three Lepidopterous Cotton Pests," United States Department of Agriculture Production Research Report 126, 1971.

(8) Fisher, R. W., Menzies, D. R., Herne, D. C., and Chida, M., "Parameters of Dicofol Spray Deposit in Relation to Mortality of European Red Mite," Journal of Economic Entomology, Vol. 67, 1974, pp. 124-126.

(9) Smith, D. B., Burt, E. C., and Lloyd, E. P., "Selection of Optimum Spray-Droplet Sizes for Boll Weevil and Drift Control," Journal of Economic Entomology, Vol. 68, 1975, pp. 415-417.

(10) Fisher, R. W. and Menzies, D. R., "Effect of Spray Droplet Density and Exposure Time on he Immobilization of Newly-Hatched Oriental Fruit Moth Larvae," Journal of Economic Entomology, Vol. 68, 1976, pp. 438-440.

(11) Jimenez, E., Roth, L. O., and Young, J. H., "Droplet Size and Spray Volume Influence on Control of the Bollworm," Journal of Economic Entomology, Vol. 69, 1976, pp. 327-329.

(12) Smith, D. B., Hostetter, D. L., and Ignoffo, C. M., "Laboratory Performance Specifications for a Bacterial (Bacillus thuringiensis) and a Viral (Baculovirus heliothis) Insecticide," Journal of Economic Entomology, Vol. 70, 1977, pp. 437-441.

(13) Smith, D. B., Hostetter, D. L., and Ignoffo, C. M., "Nozzle Size-Pressure and Concentration Combinations of Heliothis zea Control with an Aqueous Suspension of Polyvinyl Alcohol and Baculovirus heliothis," Journal of Economic Entomology, Vol. 72, 1979, pp. 920-923.

(14) Himel, C. M., "The Fluorescent Particle Spray Droplet Tracer Method," Journal of Economic Entomology, Vol. 62, 1969, pp. 912-916.

(15) Smith, D. B., Hostetter, D. L., and Ignoffo, C. M., "Ground Spray Equipment for Applying Bacillus thuringiensis Suspension on Soybeans," Journal of Economic Entomology, Vol. 70, 1977, pp. 633-637.

(16) Scott, W. P., Smith, D. B., and Lloyd, E. P., "Direct and Residual Kill of the Boll Weevil with ULV Sprays and Azinphosmethyl," Journal of Economic Entomology, Vol. 67, 1974, pp. 408-410.

(17) Wofford, J. T., "The Effects of Spray Characteristics of Vegetable Oil and Water Sprays of Permethrin on Mortality of Heliothis spp. Larvae," M.S. Thesis, Mississippi State University, Mississippi State, 1985.

(18) MacQuillan, M. J., Sucksoong, V., and Hooper, G. H. S., "Surfactant Quantity and Deposit Particle Size Influence on Direct Spray and Residual Deposit Toxicity of Chlorpyrifos to Heliothis punctigera," Journal of Economic Entomology, Vol. 69, 1976, pp. 492-493.

(19) Head, R. B., "Cotton Insect Control Guide," Publication 343, Mississippi Cooperative Extension Service, Mississippi State University, Mississippi State, 1989.

(20) Dixon, W. J., Brown, M. B., Engelman, L., Frane, J. W., Hall, M. A., Jennrich, R. I., and Toporak, J. D., "Biomedical Computer Programs (BMDP, P-Series)," University of California, Los Angeles, 1981.

(21) Finney, D. J., "Probit Analysis, 3rd Edition," Cambridge University Press, 1971.

(22) Hall, F. R., and Reichard, D. L., "Effects of Spray Droplet Size, Dosage and Solution Per Ha Rates on Mortality of Two-Spotted Spider Mite," Journal of Economic Entomology, Vol. 71, 1978, pp. 279-282.

Frank A. Manthey, Edward F. Group, Jr., and John D. Nalewaja

CUTICULAR WAX SOLUBILITY AND LEAF-CELL MEMBRANE PERMEABILITY OF
PETROLEUM SOLVENTS

REFERENCE: Manthey, F. A. , Group, E. F. Jr., and Nalewaja, J. D., "Cuticular Wax Solubility and Leaf-Cell Membrane Permeability of Petroleum Solvents", Pesticide Formulations and Application Systems: 10th Volume, ASTM STP 1078, L. E. Bode, J. L. Hazen, and D. G. Chasin, Eds., American Society for Testing and Materials, Philadelphia, 1990.

ABSTRACT: Information on petroleum solvent interaction with cuticular wax and with leaf-cell membranes may be useful in selecting the most effective solvents for pesticide formulation. Experiments were conducted with corn and green foxtail to determine petroleum solvent recoil from cuticular leaf wax, relative solvency for cuticular wax, leaf-cell membrane permeability as influenced by petroleum solvents, and fluazifop-P absorption, translocation and phytotoxicity when applied with various petroleum solvents. The solvent recoil test procedure was not reproducible. The relative solvency of petroleum solvents for cuticular leaf wax tended to be similar for both corn and green foxtail. Aromatic 100, Aromatic 200, EXXSOL D 80, NORPAR 15, and VARSOL 1 had greater solvency than ISOPAR L, ISOPAR M, ISOPAR V, ORCHEX 796, and distilled water for leaf wax. Correlations based on solvent structure are made. Solvency for cuticular wax correlated positively with leaf-cell membrane permeability. Both solvency for leaf wax and leaf-cell permeability correlated positively with solvent phytotoxicity. Fluazifop-P did not affect petroleum solvent phytotoxicity. Fluazifop-P had a negligible affect on petroleum solvent solubility for cuticular leaf wax but did reduce solvent volatility. Absorption and translocation of ^{14}C-fluazifop in both corn and green foxtail were not greatly affected by the solvents tested. The solvent which enhanced fluazifop-P phytotoxicity the most was different for corn and green foxtail. Solvents tested differed in relative wax solubility and in their affect on leaf-cell membrane permeability, but no obvious relationship occurred among these factors and the absorption, translocation and phytotoxicity of fluazifop-P.

KEYWORDS: carrier, corn, fluazifop-P, green foxtail, herbicide, adjuvancy, phytotoxicity

Frank A. Manthey, Research Scientist, is located at North Dakota State University, Crop and Weed Science Department, as is John D. Nalewaja, Professor; P. O. Box 5051, Fargo, ND 58105. Edward F. Group, Jr., Senior Staff Chemist, is with Exxon Chemical Company, P. O. Box 5200, Baytown, Texas 77522.

Many crop protection pesticides must be absorbed into plant cells to be effective. Foliar absorption involves the passage of a pesticide across the leaf cuticle, the cell wall, and the cell membrane. The outer surface of the cuticle is covered with epicuticular wax. The cuticle contains cutin, an insoluble polymer composed of crosslinked hydroxy-fatty acids [1]. Associated with the cutin polymer are wax-like lipid and nonlipid components such as polysaccharides, polypeptides, and phenolics. Cuticular waxes and cell membrane are generally nonpolar while cutin and cell wall are polar. The ratio of polar to non-polar components varies with species and environment. Thus, a pesticide must penetrate both polar and non-polar barriers before being absorbed into plant cells.

Solvents used for pesticide formulation can serve an important function in the enhancement of spray retention, area of contact and/or stomatal penetration and cuticular penetration of a pesticide [2-5]. Solvents used as formulants, or as spray adjuvants or carriers, can solubilize cuticular waxes and thus enhance pesticide absorption [6-8].

Solvents may injure plant cell membranes which may reduce pesticide absorption into and/or translocation out of the cells, thus reducing efficacy [5]. Solvents that injure leaf tissue tend to be those that more readily dissolve cuticular waxes; however, not all solvents that readily dissolve cuticular wax cause leaf injury [8].

Research was conducted with corn and green foxtail to determine solvent recoil from leaf wax, relative solubility for cuticular wax, leaf-cell membrane permeability by petroleum solvents, and absorption, translocation, and phytotoxicity of fluazifop-P applied with petroleum solvents.

EXPERIMENTAL PROCEDURES

General Procedures

Corn was seeded in 1.0 L and green foxtail in 0.5 L plastic pots containing a 50:50 mix of peat and sandy loam soil. Corn was thinned to three plants per pot and green foxtail was thinned to seven plants per pot one week after emergence. Plants were watered and fertilized as needed for optimum growth. The natural daylength was supplemented for a 16-h photoperiod with metal-halide lamps with a plant level intensity of 450 $uE \cdot m^{-2} \cdot s^{-1}$. The greenhouse was maintained at 20 C at night and 30 C during the day with a 5 C variation.

Solvents were applied at 9.4 L/ha (1 gpa) using an air brush sprayer. The air brush sprayer was calibrated for each solvent by varying orifice size and air pressure. The soil was covered with vermiculite before treatment (and removed after treatment) to prevent the solvents and/or herbicide from contacting the soil.

Based on results obtained in previous work [8], a smaller group of representative solvents and oils were chosen for further testing; they are listed and described in the following LEGEND.

LEGEND
DESCRIPTION OF PRODUCTS TESTED[a]

TRADE NAME	DESCRIPTION/TYPE	TYPICAL DISTILLATION RANGE, °C
Solvents and Oils[b]		
Aromatic 100	C-9 Alkylbenzenes (aromatic)	156-169
Aromatic 200	C-11/C-12 Alkylnaphthalenes	229-271
EXXSOL D 80	C-12/C-13 Dearomatized Aliphatic	204-234
ISOPAR L	C-12/C-13 Isoparaffinic	191-207
ISOPAR M	C-14 average Isoparaffinic	223-249
ISOPAR V	C-17 average Isoparaffinic	276-308
NORPAR 15	C-15 average normal paraffinic	249-276
ORCHEX 796	Paraffinic petroleum oil (C-23)	385[c]
VARSOL 1	Mineral spirits (C-10/C-11)-- 15-20% aromatics content	157-197

[a]For further details on these solvents and oils and their composition and properties, see ASTM STP 980, Pesticide Formulations and Application Systems 8th Volume; An Overview--Solvents for Agricultural Chemicals, by M.R. Krenek and W.H. Rohde (1988).

[b]All supplied by Exxon Company U.S.A. and Exxon Chemical Company. EXXSOL, ISOPAR, NORPAR, VARSOL and ORCHEX are registered trademarks of Exxon Corporation.

[c]50% point by simulated (GC) distillation.

Solvent Recoil and Relative Wax Solubility

Wax from the leaves of corn and green foxtail was removed by a 10 sec chloroform dip [1]. The chloroform solution was filtered to remove debris. Acetone was added to precipitate the wax, while non-wax components and pigmentation remained dissolved in the acetone [9]. After filtration, the precipitate was dissolved in chloroform to a concentration of 15 mg/ml. One 50 ul droplet of chloroform-wax solution was placed on a glass microscope slide and allowed to dry for at least 24 h.

As in earlier work [8], solvent recoil and relative solubility
of wax in solvents were determined by placing a 3 ul droplet of
solvent on the wax. After 2.5 min, 1 h, or 24 h, the slide was
tilted to allow the solvent to drain off the wax. The slide was
tapped to aid drainage of the more viscous solvents and oils. A
solvent recoil number was assigned according to how the solvent
reacted to the wax. The rating scale ranged from 1 to 5 where 1=no
recoil or contraction of the solvent from the wax; 2= slight recoil
of the solvent from the wax; 3= moderate recoil, 4= moderately
severe recoil, and 5= severe recoil of the solvent from the wax.
Next, the wax was firmly wiped perpendicular to the direction of the
solvent drainage and a relative wax solubility number was assigned.
The rating scale ranged from 1 to 9 where 1= no solubility; 3=
smeared circle where the solvent droplet was placed; 5= clear circle
where the solvent droplet was placed; 7= smeared oblong where the
solvent droplet was placed and drained off; and 9= clear oblong
where the solvent droplet was placed and drained off. The
experiments were conducted as a split-plot design, where the whole
plots were time of the ratings, and sub-plots were solvent
treatments. The experiments had six replications. Means were
separated using a protected least significant difference (LSD) test.

Leaf-Cell Membrane Permeability

Corn was thinned to three plants per pot and green foxtail was
thinned to four plants per pot. Treatments were applied at 9.4 L/ha
using an air brush sprayer, to green foxtail in the 5-leaf stage,
and to corn in the 4-leaf stage. Twenty discs, five from the fourth
leaf of each green foxtail plant in a pot, and thirty discs, ten
from the third leaf of each corn plant in a pot, were punched 2.5
min and 24 h after treatment. All discs were 0.6 mm in diameter.
Care was taken to avoid the midrib of the corn leaf. The leaf discs
were incubated in 20 ml of distilled water at 20 C and shaken at 90
rpm for 6 h. Electrolyte leakage was measured with a conductivity
bridge using a conductivity cell (k=1.0). Corn and green foxtail
were separate experiments. The experiments were conducted as a
split-plot design, where the whole plots were time after treatment
and sub-plots were solvent treatments. Each experiment was
conducted 12 times. The data were combined over experiment
repetitions. Means were separated using a protected least
significant difference (LSD) test. Procedural experiments indicated
that the solvents and solvent-fluazifop-P combinations themselves
did not affect the electroconductivity readings of the incubation
solutions.

Efficacy Experiments

Treatments were applied to 5-leaf green foxtail and 3 1/2-leaf
corn. Solvents were applied at 9.4 L/ha total spray volume and
fluazifop-P (commercial formulation, 1 lb ai/gal) was applied at 20
g ai/ha to corn and at 25 g/ha to green foxtail, using solvents as
spray carriers. The spray solution contained solvent and
fluazifop-P in a ratio of 7/1 v/v, respectively. Injury ratings,

where 0=no injury and 100= complete kill, were taken 12 to 24 h, and 14 days after treatment. All greenhouse experiments were a randomized complete block design with four replications and were repeated. Each plant species was conducted as a separate experiment. The data were combined over experiment repetitions. Means were separated using a protected least significant difference (LSD) test.

[14]C-Fluazifop Absorption and Translocation

Corn and green foxtail were sown in soil contained in 0.5 L plastic pots in the greenhouse. Plants at the 1-leaf stage were thinned to three per pot and moved to controlled environmental chambers at 23 C, 40 to 50% relative humidity, and a 16 h photoperiod with an intensity of 250 $uE.m^{-2}.s^{-1}$. Plants were thinned to one per pot one day before treatment.

Treatments were applied and spread to cover 1 cm^2 area midway between the tip and base of the fourth leaf of green foxtail in the 5-leaf stage, and of the second leaf of corn in the 2 1/2-leaf stage.

Details of the treatment procedures were as previously described [10]. The ^{14}C-fluazifop (21.5 mCi/mmole specific activity) was applied at 0.013 uCi. The formulated and ^{14}C-fluazifop were applied in an amount to equal 140 g ai/ha using the treated leaf area as the basis to establish a hectare. The solvents were applied at 9.4 L/ha.

The plants were removed from the soil and the roots washed in tapwater 24 h after treatment with ^{14}C-fluazifop. Each plant was sectioned into treated area, remaining treated leaf, and remaining shoot and root. The treated area was washed by dipping 10 times in 15 ml scintillation fluid, 1:1 (v/v) toluene:ethanol with 5 g PPO and 0.5 g POPOP scintillation fluors per liter. All plants were dried in a forced air drier at 70 C for 48 h. The dried plant parts were combusted in a biological material oxidizer; the [14]CO_2 was collected in 15 ml scintillation fluid (10:7:3 v/v/v) toluene:2-methoxyethanol:ethanolamine plus 15 g PPO and 1.5 g POPOP/L; and assayed for radioactivity in a scintillation spectrometer. The total amount of ^{14}C absorbed and translocated out of the leaf was determined following appropriate quench correction.

Disintegrations per min were expressed as percentage of the applied [14]C. Each experiment had a completely random design with treatments repeated four times. Each experiment was conducted twice. The data were combined over experiment repetitions and an analysis of variance was conducted with each variable. Means were separated using a protected least significant difference (LSD) test.

RESULTS AND DISCUSSION

Solvent recoil ratings varied greatly among experiments (data not presented). The recoil test procedure described in this paper is too subjective to be reliably reproduced. Thus, this procedure does not provide adequate means for assessing solvent droplet interaction with leaf wax.

Relative wax solubility numbers for solvents tended to be similar for corn and green foxtail wax (Table 1). Aromatic 100, Aromatic 200, EXXSOL D 80, NORPAR 15, and VARSOL 1 readily dissolved corn and green foxtail cuticular leaf wax. ISOPAR L readily dissolved green foxtail leaf wax but not corn leaf wax. Relative wax solubility numbers for solvents varied depending on the time of observation. For example, relative wax solubility ratings for Aromatic 100 and VARSOL 1 decreased with time while for ISOPAR V and ORCHEX 796 the ratings increased with time (from 2.5 min to 24 h) for both corn and green foxtail wax. Those solvents whose ratings decreased at 24 h had volatilized and redeposited leaf wax onto the glass slide.

Relative wax solubility ratings at 2.5 min correlated positively with solvent toxicity to corn (r=0.85) and green foxtail (r=0.67). Aromatic 100, Aromatic 200, and VARSOL 1 dissolved leaf wax quickly and were phytotoxic; however, EXXSOL D 80, and NORPAR 15 dissolved leaf wax quickly but were not phytotoxic (Table 1). Relative wax solubility ratings at 1 and 24 h did not correlate with solvent phytotoxicity. Solvents which dissolved wax slowly were not phytotoxic even though they had relative wax solubility ratings at 1 h and/or 24 h which were similar to the phytotoxic solvents at 2.5 min. Thus, solvents which readily dissolve wax are more likely to cause leaf injury than those that slowly dissolve leaf wax. However, not all solvents which readily dissolve leaf wax were phytotoxic.

The structural make-up of solvents tested (see LEGEND) correlated reasonably well with results obtained. Solvents with high aromatics content generally had good solvency for cuticular leaf wax and were more phytotoxic. EXXSOL D 80, comprised of roughly equal parts of isoparaffinic and cycloaliphatic (naphthenic) structures, had better initial (2.5 min) solvency for both corn and green foxtail leaf wax compared with ISOPAR grades, yet was non-phytotoxic. Comparing ISOPAR L and ISOPAR M (very similar isoparaffinic structures), volatility of the solvent came into play; ISOPAR L had better initial solvency for green foxtail but lacked the staying power of ISOPAR M which was higher in molecular weight (lower volatility). Also evident from Table 1 was the ability of NORPAR 15, a linear C-15 paraffin with wax-like properties, to quickly dissolve leaf wax, yet was non-phytotoxic.

TABLE 1:   Ratings of solvency for corn and green foxtail leaf
            wax relative to plant injury

Solvent	Corn			Injury 24 h	Green foxtail			Injury 24 h
	2.5min	1 h	24 h	24 h	2.5min	1 h	24 h	24 h
	--- (RWS)[a] --			%	--- RWS) ----			%
Aromatic 100	6.8	5.0	1.0	33	4.5	4.5	3.5	75
Aromatic 200	4.8	5.2	5.5	34	4.7	3.5	4.8	68
EXXSOL D 80	3.5	5.3	4.5	0	4.2	4.0	2.0	1
ISOPAR L	1.5	5.0	1.0	0	3.3	3.3	1.2	0
ISOPAR M	1.0	4.2	4.7	0	1.5	3.7	4.2	1
ISOPAR V	1.0	1.5	3.5	1	1.0	2.0	4.2	1
NORPAR 15	3.8	4.7	4.8	0	3.5	3.7	4.7	0
ORCHEX 796	1.0	1.8	2.8	0	1.0	2.5	4.7	8
VARSOL 1	5.2	2.3	1.0	28	4.2	4.2	1.8	59
Distilled water	1.0	1.0	1.0	0	1.0	1.0	1.0	0
LSD (0.05)	----- 0.8 ----			7	----- 0.8 ----			9

[a]RWS is relative wax solubility rating. Evaluation scale 1.0=No
solubility; 3.0=Smeared circle where solvent droplet was placed;
5.0=Clear circle where solvent droplet was placed; 7.0=Smeared
oblong where solvent droplet was placed and drained off; 9.0=Clear
oblong where solvent droplet was placed and drained off.

    Fluazifop-P as a formulated herbicide controls grass species in
various broadleaf crops. Fluazifop-P applied with the solvents at
the concentration used in the efficacy experiments had little to no
affect on the relative wax solubility numbers for the solvents in
corn and green foxtail wax at 2.5 min (Table 2). Fluazifop-P in
distilled water tended to solubilize corn wax after 24 h and green
foxtail wax at 1 and 24 h after treatment, presumably due to
presence of formulants in the herbicide. Fluazifop-P appeared to
reduce the volatility of Aromatic 100, EXXSOL D 80, and distilled
water, as evidenced by a wet spot after 24 h, but the same solvents
evaporated in 1 to 24 h when applied alone.   Thus, fluazifop-P
increased the relative wax solubility ratings after 24 h for the
relatively volatile Aromatic 100, EXXSOL D 80 and distilled water.
These results indicate that a formulated pesticide can influence the
outcome of experiments comparing different solvent types.

TABLE 2: Ratings of solvency for corn and green foxtail leaf wax, as influenced by fluazifop-P.

Solvent	Fluaz-ifop-P	Corn			Green foxtail		
		2.5min	1 h	24 h	2.5min	1 h	24 h
		--- (RWS[a]) ---			------ (RWS) -----		
Aromatic 100	No	4.8	4.7	1.0	4.8	4.8	3.0
	Yes[b]	5.0	4.8	4.3	4.5	4.7	4.8
EXXSOL D 80	No	4.0	4.8	1.0	2.8	4.7	1.2
	Yes	3.3	4.5	2.5	2.2	3.5	3.8
ISOPAR V	No	1.0	1.2	3.5	1.0	2.3	4.0
	Yes	1.0	2.2	3.2	1.0	2.5	4.3
ORCHEX 796	No	1.0	1.5	2.8	1.0	2.5	4.8
	Yes	1.0	2.0	2.7	1.0	2.0	4.0
Distilled water	No	1.0	1.0	1.0	1.0	1.0	1.0
	Yes	1.0	1.2	1.5	1.0	4.3	3.8
LSD (0.05)		----- 0.6 ------			----- 0.7 -----		

[a]RWS is relative wax solubility rating. Evaluation scale 1.0=No solubility; 3.0=Smeared circle where solvent droplet was placed; 5.0=Clear circle where solvent droplet was placed; 7.0=Smeared oblong where solvent droplet was placed and drained off; 9.0=Clear oblong where solvent droplet was placed and drained off.

[b]Droplet contained 87.5% solvent and 12.5% fluazifop-P (commercial product).

Electroconductivity of the solution in which leaf discs treated with the solvents were incubated was used to measure the effect of solvents on leaf-cell membrane permeability. Injured leaf cells leaked electrolytes into the incubation solution. Thus, electroconductivity of the incubation solution increased with increased leaf-cell permeability.

Aromatic 100, Aromatic 200, and VARSOL 1 increased corn leaf-cell permeability 2.5 min after treatment compared to the permeability of corn leaf-cells treated with distilled water (Table 3). This result is in line with presence of aromatic components in these three solvents. The other solvents when compared to distilled water had a negligible effect on corn leaf-cell permeability 2.5 min or 24 h after treatment. Leakage of electrolytes from corn leaf cells was less for 24 h than for 2.5 min after treatment with Aromatic 100, Aromatic 200, or VARSOL 1, although significant corn leaf injury was apparent 24 h after treatment (Table 1). The corn leaf injury would be expected from the cell injury indicated by electrolyte leakage after treatment. The lower electroconductivity values at 24 h compared to 2.5 min after treatment may indicate electrolyte adsorption to dead tissue.

Green foxtail leaf-cell permeability was greater at both 2.5 min and 24 h after treatment with Aromatic 100, Aromatic 200, EXXSOL D 80, ISOPAR L, or VARSOL 1 than with distilled water (Table 3). However, only Aromatic 100, Aromatic 200, and VARSOL 1 injured green foxtail 24 h after treatment (Table 1). The electroconductivity values for Aromatic 100, Aromatic 200, and VARSOL 1 were greater than for EXXSOL D 80 and ISOPAR L, at 2.5 min and 24h after treatment, which indicated a cell permeability threshold may exist above which visible injury occurs. The cell permeability threshold probably varies with plant species.

Electroconductivity values at 2.5 min correlated positively with relative wax solubility ratings for the solvents 2.5 min after treatment; for corn r=0.61, and for green foxtail r=0.89. Thus, leaf-cell permeability increased with increased solvent solubility for leaf wax. Electroconductivity values at 24 h did not correlate with relative wax solubility ratings at 24 h for either corn or green foxtail. Several solvents dissolved leaf wax within 24 h but did not cause electrolyte leakage. The solvents which dissolved leaf wax, but did not enhance electrolyte leakage, apparently were unable to cross the cell wall quickly enough or in sufficient quantity to alter leaf-cell permeability, and/or did not affect cell membrane permeability.

Electroconductivity values at 2.5 min and 24 h correlated positively with solvent phytotoxicity to corn (r=(0.82 and 0.79) and green foxtail (r=0.86 and 0.93). Thus, solvent phytotoxicity increased with increased leaf-cell permeability. Solvent phytotoxicity was correlated positively with both relative wax solubility ratings and leaf-cell permeability. However, two exceptions, EXXSOL D 80 and ISOPAR L, readily dissolved green foxtail wax and increased green foxtail cell membrane permeability at 2.5 min but did not cause any apparent injury (Tables 1 and 3).

Corn leaf-cell membrane permeability 2.5 min and 24 h after treatment was less with than without fluazifop-P in Aromatic 100 (Table 4). However, leaf-cell permeability 24 h after treatment was greater with than without fluazifop-P in distilled water with corn leaves and in Aromatic 100, EXXSOL D 80 and distilled water with green foxtail leaves. The effect of fluazifop-P on cell permeability when applied with Aromatic 100, EXXSOL D 80, and distilled water was most likely a direct effect of fluazifop-P and/or its formulants on cell membrane permeability.

TABLE 3:   Leakage of electrolytes from corn and green foxtail leaf
tissue treated with petroleum solvents.

Solvent	Corn		Green foxtail	
	2.5 min	24 h	2.5 min	24 h
	--- (Electroconductivity, umhos/cm) ---			
Aromatic 100	60	50	137	118
Aromatic 200	113	73	146	138
EXXSOL D 80	45	35	103	78
ISOPAR L	47	33	86	47
ISOPAR M	47	30	41	40
ISOPAR V	42	34	41	38
NORPAR 15	46	31	42	38
ORCHEX 796	40	40	41	40
VARSOL 1	169	42	126	102
Distilled water	36	36	38	38
LSD (0.05)	---- 11 ----		----- 19 -----	

TABLE 4.   Electrolyte leakage from corn and green foxtail treated
with petroleum solvents, as influenced by fluazifop-P.

| | Fluaz-ifop-P | -- (Electroconductivity, umhos/cm) -- | | | |
| | | Corn | | Green foxtail | |
		2.5 min	24 h	2.5 min	24 h
Aromatic 100	No	66	58	109	102
	Yes[a]	50	44	115	134
EXXSOL D 80	No	39	33	66	27
	Yes	39	39	60	62
ISOPAR V	No	33	30	31	29
	Yes	33	33	33	37
ORCHEX 796	No	34	35	31	29
	Yes	34	38	31	35
Distilled water	No	32	32	27	29
	Yes	36	40	32	59
LSD (0.05)		---- 7 -----		----- 20 ----	

[a]Spray solution applied at 9.4 L/ha; contained 87.5% solvent and
12.5% fluazifop-P (commercial product)

Aromatic 100 was the only solvent to cause substantial injury to corn or green foxtail 1 DAT (day after treatment) (Table 5). Corn and green foxtail injury 1 DAT was less when Aromatic 100 was applied with than without fluazifop-P. Rapid injury from Aromatic 100 indicated direct phytotoxicity to the plant cells. Little to no injury to corn and green foxtail occurred with the other solvents applied alone or with fluazifop-P 1 DAT.

Solvents applied without fluazifop-P caused little injury to either corn or green foxtail 14 DAT, except for Aromatic 100. Fluazifop-P caused 82 to 89% injury to corn and 65 to 83% injury to green foxtail, depending on the solvent (Table 5). Injury to corn 14 DAT tended to be greater when fluazifop-P was applied in a distilled water spray carrier than in an Aromatic 100, ISOPAR V, or ORCHEX 796 spray carrier but differences were small. Injury to green foxtail was greater when fluazifop-P was applied in an Aromatic 100, EXXSOL D 80, or distilled water carrier than an ISOPAR V or ORCHEX 796 spray carrier.

The air brush sprayer used in this research delivered a low volume application (9.4 L/ha) with small spray droplets. Small concentrated spray droplets containing fluazifop were more phytotoxic than large, less concentrated droplets when applied in water carrier [11]. Small spray droplets may result in greater spray retention; hence, more fluazifop-P on the plant. Fluazifop-P contained various formulants which dissolved leaf wax of corn and green foxtail (Table 2). The concentration of formulants in the water-fluazifop-P spray droplets might have been adequate for the absorption of fluazifop-P.

Absorption of ^{14}C by corn was similar regardless of solvent applied with ^{14}C-fluazifop. Translocation of ^{14}C to the shoot and root was generally least when ^{14}C-fluazifop was applied with Aromatic 100, probably due to localized injury caused by the Aromatic 100. Absorption of ^{14}C by green foxtail was greatest when ^{14}C-fluazifop was applied with ISOPAR V and ORCHEX 796 and the least when applied with EXXSOL D 80. However, ^{14}C translocation in green foxtail to shoot and root was similar regardless of solvent applied with ^{14}C-fluazifop. Thus increased ^{14}C-fluazifop absorption did not result in increased translocation.

Previous research had indicated that fluazifop absorption and translocation by green foxtail was enhanced by an oil adjuvant [12]. Fluazifop-P used in the previous research [8] was a 480 g ai/L (4 lb ai/gal) formulation whereas we used fluazifop-P which was a 120 g ai/L (1 lb ai/gal) formulation. Therefore, treatments in this study contained more formulants than in previous work, which may account for the increased absorption. The relatively high absorption of fluazifop-P when applied in water is evidenced by higher leaf wax solubility, leaf-cell permeability and phytotoxicity of distilled water containing fluazifop-P, for both corn and green foxtail.

TABLE 5: Fluazifop-P injury and absorption and translocation by corn and green foxtail as influenced by petroleum solvents.

Solvent	Fluaz-ifop-P	Corn DAT[a] 1	14	Absorb	Shoot/root	Green foxtail DAT 1	14	Absorb	Shoot/root
		% inj		% of applied		% inj		% of applied	
Aromatic 100	No	33	37	...	...	75	59	...	...
	Yes	3	82	71.1	5.7	66	83	73.5	3.5
EXXSOL D 80	No	0	6	...	...	1	1	...	...
	Yes	0	86	73.1	7.0	1	81	67.8	3.3
ISOPAR V	No	1	6	...	...	1	0	...	...
	Yes	0	83	71.1	8.5	3	65	79.9	4.3
ORCHEX 796	No	0	8	...	...	8	1	...	...
	Yes	0	84	75.7	8.0	12	74	77.2	3.7
Distilled water	No	0	9	...	...	0	0	...	...
	Yes	0	89	69.8	8.5	1	80	73.2	3.7
LSD (0.05)		4	5	NS	1.9	5	5	5.7	0.8

[a]DAT is days after treatment

ACKNOWLEDGEMENTS: The authors acknowledge the technical assistance of Wanda Eastman, Patricia Evans, and Ronald Roach.

References

[1] Martin, J.T. and Juniper, B.E. Eds., The Cuticle of Plants, St. Martin's Press. New York, 1970.

[2] Baker, D.A., Hunt, G.M., and Stevens, P.J.G., "Studies of Plant Cuticle and Spray Droplet Interactions: A Fresh Approach", Pesticide Science, Vol. 14, 1983, pp. 645-658.

[3] Hess, F.D., "Herbicide Absorption and Translocation and Their Relationship to Plant Tolerance and Susceptibility", in Weed Physiology, Vol. II. Herbicide Physiology, CRC Press, Boca Raton, FL, 1985, pp. 191-214.

[4] Stevens. P.J.G. and Baker, E.A., "Factors Affecting the Foliar Absorption and Redistribution of Pesticides: I. Properties of Leaf Surfaces and Their Interactions with Spray Droplets", Pesticide Science, Vol, 19, 1987, pp. 265-281.

[5] Crafts, A.S., "Herbicide Behaviour in the Plant", in The Physiology and Biochemistry of Herbicides, L.J. Audus, Ed., Academic Press, New York, 1964, pp. 75-110.

[6] Price, C.E., "A Review of the Factors Influencing the Penetration of Pesticides through Plant Leaves", in The Plant Cuticle, Academic Press, New York, 1982, pp. 237-252.

[7] Saunders, R.K. and Lonnecker, W.M., "Physiological Aspects of Using Nonphytotoxic Oils with Herbicides", Proceedings North Central Weed Control Conference, Vol. 21, 1967, pp. 62-63.

[8] Manthey, F.A., Nalewaja, J.D., Group, E.F., Jr., and Krenek, M.R., "Epicuticular Wax Solubility in Petroleum Solvents Relative to Herbicide Phytotoxicity", 9th Symposium on Pesticide Formulations and Application Systems "International Aspects", ASTM STP 1036, James L. Hazen, David A. Hovde, Eds., American Society for Testing and Materials, Philadelphia, 1989, in press.

[9] Kurtz, E.B., Jr., "The Relation of the Characteristics and Yield of Wax to Plant Age, "Plant Physiology, Vol. 25, 1950, pp. 269-278.

[10] Nalewaja, J.D. and G.A. Skrzypczak, "Absorption and Translocation of Fluazifop with Additives," Weed Science. Vol 34, 1986, pp. 572-576.

[11] Buhler, D. D., and O. C. Burnside, "Effect of Application Factors on Postemergence Phytotoxicity of Fluazifop-butyl, Haloxyfop-methyl, and Sethoxydim," Weed Science. Vol. 32, 1984, pp. 574-583.

[12] Grafstrom, L.D., Jr. and Nalewaja, J.D. "Uptake and Translocation of Fluazifop in Green Foxtail (Setaria viridis)", Weed Science. Vol 36, 1988, pp. 153-158

Masoud Salyani, James P. Syvertsen, and Joseph L. Knapp

EFFECTS OF SPRAY OILS ON TEMPERATURE, NET GAS EXCHANGE, AND
PHYTOTOXICITY OF CITRUS LEAVES

REFERENCE: Salyani, M., Syvertsen, J. P., and Knapp, J. L.,
"Effects of Spray Oils on Temperature, Net Gas Exchange, and
Phytotoxicity of Citrus Leaves," Pesticide Formulations and
Application Systems: 10th Volume, ASTM STP 1078, L. E. Bode,
J. L. Hazen, and D. G. Chasin, Eds., American Society for
Testing and Materials, Philadelphia, 1990.

ABSTRACT: Three types of horticultural spray oils (Sunspray
7N, Sunspray 9N, and VOLCK Supreme Spray) were used to
prepare spray mixtures of 1, 2, and 4% (V/V) oil plus
emulsifier in water. The 50% distillation temperatures (50%
DT) of the oils were 224, 235, and 247°C (435, 455, and
476°F), respectively. The mixtures were applied to the
fully expanded young leaves of 2-year-old containerized
'Valencia' orange trees in 2 tests. In the first test,
leaves on different plants were dipped in agitated oil
mixtures and after one day, under sunny conditions, their
leaf temperature (T_1, infrared thermometer) and stomatal
conductance (G_s, diffusion porometer) were measured. There
was no significant effect of oil DT on T_1 or G_s. However,
averaged over all DTs of oil, the 4 and 2% rates
significantly decreased G_s below that of the 1% rate. In
the second test, the 1 and 4% oil mixtures and an emulsifier
only solution were uniformly sprayed on both surfaces of the
leaves with a hand atomizer. Three to 5 days later,
stomatal conductance, CO_2 assimilation (A), and dark
respiration (R) of sprayed leaves were determined, using an
open gas exchange system in the laboratory. Neither oil DT
nor oil concentration had a significant effect on G_s or A;
however, R was highest with Sunspray 7N-1% and lowest with
Sunspray 9N-4% treatments. In both tests, no visible
phytotoxicity was observed after 3 months under glasshouse
conditions.

KEYWORDS: petroleum spray oil, citrus, phytotoxicity,
stomatal conductance, CO_2 assimilation, dark respiration

Dr. Salyani is Assistant Professor of Agricultural Engineering,
Dr. Syvertsen is Professor of Plant Physiology, and Dr. Knapp is
Professor of Entomology, at the University of Florida, IFAS, Citrus
Research and Education Center, Lake Alfred, FL 33850.

Petroleum spray oils of varying specifications are widely used in the citrus industry to control economically important arthropod pests [1,2] and greasy spot fungus [3,4] in the tree canopy. In addition to their pesticidal properties, they are relatively safe for the user, have little adverse effect on biological control agents [1], are unlikely to induce pest resistance to oil's activity, create minimal pesticidal residue problems [5], and are exempt from the United States Environmental Protection Agency's requirement for a tolerance. Agricultural spray oils are also considered to be good spreaders for other pesticides [1]. Nevertheless, improper oil or excessive use of it may develop various adverse effects (phytotoxicity) on the tree and in fruit. Phytotoxic responses of citrus such as reduced fruit yield and size [6], excessive leaf abscission and fruit drop, reduced bloom and fruit set, fruit blemishes and poor color and quality, and increased susceptibility to cold weather injury have been reported [1,5]. However, phytotoxic responses are quite variable since they depend on characteristics of the oil, oil-pesticide interaction [7], spraying time and weather conditions [6], and method of application [2].

Through utilizing different crudes and refining processes, spray oils can be produced with different hydrocarbon compositions and physical properties. However, the current superior horticultural oils use refined petroleum distillate which tends to be more paraffinic than naphthenic or aromatic [8]. The reason being that aromatics are unstable hydrocarbons that are more likely to cause phytotoxicity than other oil components. Pesticidal activity and potential phytotoxic effects of oils have been related to their hydrocarbon composition as well as to their physical properties. Oils of good quality fit the following specifications; distillation midpoints between 204 and 249°C (400-480°F), a narrow distillation 10-90% range of around 44°C (80°F) maximum (based on ASTM D-1160 performed at 10 mm (Hg), an unsulfonated residue of 92% minimum (based on ASTM D-483), an API gravity of 31 or above and pour points of -12°C (10°F) or lower [8]. Trammel and Simanton [1] found that insecticidal efficacy of spray oils against citrus red mite and Florida red scale increased as distillation midpoint temperature increased, but beyond a certain optimum range, efficacy decreased.

Oils presumably kill mites and insects by smothering them [1]. However, the precise mode of action of oil in reducing the susceptibility of citrus leaves to greasy spot infection [3] is not known but may be related to physical changes that the oil causes within the boundary layer of the leaf-air interface, or even within leaf tissues after it enters the stomata [3].

Spray oils have been reported to reduce net gas exchange of citrus leaves [9]. The mechanism of phytotoxicity has been attributed to the physical presence of oil on stomatal surfaces, thereby reducing transpirational fluxes [10], and possibly leading to increases in daytime leaf temperatures. The purpose of our experiments was to test this hypothesis by measuring leaf temperature and net gas exchange of CO_2 and H_2O vapor of citrus leaves after treating them with spray oils having 3 different distillation temperatures.

MATERIALS AND METHODS

Three horticultural spray oils with different distillation temperatures (DT), Sunspray 7N, Sunspray 9N (Sun Refining and Marketing

Co.), and VOLCK Supreme Spray (Chevron Chemical Co.) were used to prepare oil in water mixtures of 1, 2, and 4% (V/V) along with emulsifier T-MULZ-FLO (Thompson-Hayward Co.) at 0.82% of oil volume. Specifications for the 3 oils obtained by the ASTM standard methods are given in Table 1.

TABLE 1 -- Specifications of the petroleum spray oils tested.

Oil Properties	ASTM Standard Method	Spray Oils		
		Sunspray[a] 7N	Sunspray[a] 9N	VOLCK Supreme Spray[b]
Hydrocarbon Composition, %	D-2140			
Paraffins		59	65	68
Naphthenes		40	33	32
Aromatics		1	2	0
Distillation Temp. @ 10 mm Hg, °C (°F)	D-1160			
10%		206 (402)	227 (440)	231 (447)
50%		224 (435)	235 (455)	247 (476)
90%		243 (470)	260 (500)	278 (532)
10-90%, Max		44 (80)	44 (80)	47 (85)
Unsulfonated Residue, Vol. %, Min	D-483	92	92	92
API Gravity @ 15.5°C (60°F)	D-287	32.5	33.6	34.8
SUS Viscosity @ 37.8°C (100°F), s	D-2161	85	98	105
Pour Point, °C (°F)	D-97	-18 (0)	-12 (10)	-12 (10)

[a]Information supplied by Sun Refining and Marketing Company.
[b]Information supplied by Chevron Chemical Company.

To prepare the mixtures, first the required volumes of oil and emulsifier were pipetted into a beaker and mixed by a magnetic mixer (Mag-Mix; GCA/Precision Scientific) for about 5 min. Deionized water was then added to make a total volume of 500 mL and the mixture was mixed (agitated) for about 5 min. Twelve mixtures were prepared. Nine contained oil plus emulsifier (3 oils x 3 concentrations) and 3 with emulsifier only as controls. The latter had emulsifier in the amounts equivalent to those in 1, 2, and 4% oil plus emulsifier mixtures. The stability of the mixture (absence of creaming) is known to be dependent on water hardness, vigor and duration of agitation, order and rate of combining ingredients, ambient temperature, size and shape of containers, and volume and depth of emulsion [11]. Preliminary observations (using a fluorescent dye and ultra-violet light) showed a 5 min mixing to be sufficient for uniform and stable mixing. All the mixtures were prepared at the same degree of agitation, using same size stirrer and setting on the magnetic mixer. Although adding the concentrate (oil plus emulsifier) to the water could give more stable emulsion than the reverse order [11], we practiced the latter in order to expedite the test.

Test 1

The 12 mixtures were applied to the fully expanded young leaves of
containerized 'Valencia' orange trees (budded onto Carrizo citrange
rootstock). The plants were placed outside the glasshouse one week
prior to application to be acclimated to the outside weather
conditions. Ten to 15 leaves on each tree were tagged and numbered in
order to be used for two series of tests.

To apply the mixtures to the leaves, each was selected at random,
placed on the mixer, and agitated for 2 min, then the agitating mixture
was placed under leaves, a numbered leaf was dipped in the mixture for
3 s (Fig. 1), and excess solution was removed from the leaf tip with a
wiping cloth. The procedure was repeated for all solutions and
replicated 5 times. The plant containers were watered thoroughly to
avoid drought stress and left outdoors.

One day after application, midday adaxial leaf surface temperatures
were measured (on the dipped leaves held perpendicular to the incident
radiation), using an infrared thermometer (Barnes 14-220, sensitivity =
$\pm$ 0.2°C at 20°C) and 24 gauge copper-constantan thermocouples clipped
to abaxial leaf surfaces. Stomatal resistance (inverse of stomatal
conductance, G_s) was measured with a Delta-T (AP3) diffusion porometer.

Test 2

Six mixtures of the 3 oils (plus emulsifier) at 1 and 4% and an
emulsifier only solution (with emulsifier concentration equivalent to
that in 1% oil mixture) were used in the second test. These 7 mixtures
were applied with a hand atomizer (Fig. 2) to both surfaces of the
randomly selected leaves and replicated 5 times. Three to 5 days
later, net gas exchange of CO_2 and H_2O vapor was evaluated in the
laboratory using single intact leaves and an open system [12]. The
system incorporated a well-stirred temperature controlled cuvette [13].
All net CO_2 assimilation (A) measurements were made under saturating
light conditions, 800 μmol m^{-2} s^{-1} PAR (photosynthetically active
radiation), leaf temperatures of 24 $\pm$ 1°C and vapor pressure deficits
of 14 $\pm$ 4 mb. Stomatal conductances were calculated from transpiration
rates [14]. After measurements in both tests, the trees were moved
into the glasshouse and observed for potential leaf yellowing or leaf
drop.

Field Test

A field test was also carried out to assess the phytotoxic
responses of mature grapefruit trees to oil application in the field
conditions. Sunspray 7N and Sunspray 9N oil (plus emulsifier) mixtures
were applied at 1, 2, and 4% (at volume rates of 8,400, 4,700, and
2,350 L/ha, respectively) by an airblast sprayer. The 6 oil treatments
and an unsprayed check were assigned in a randomized block design and
replicated 4 times. Number of leaves dropped from trees were counted
at 1 and 2 weeks post-treatment. These data were used as an indicator
of the phytotoxicity.

RESULTS AND DISCUSSION

Visual observations on oil distribution (in mixture container and
on leaf surfaces) revealed that oil mass in the mixture breaks into

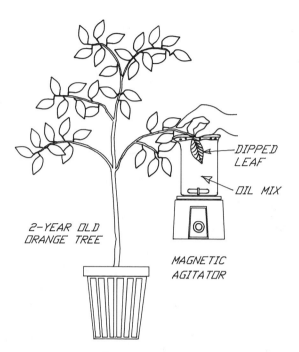

Fig. 1 -- Orange leaves dipped in oil mixture.

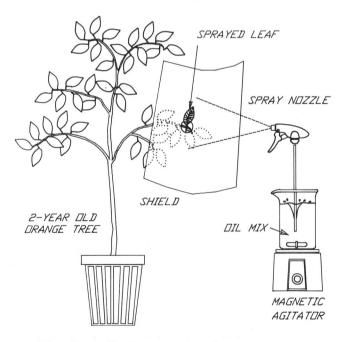

Fig. 2 -- Oil mixture sprayed on orange leaves.

small suspended droplets that move separately in the mixture. We found that the uniformity of oil droplet distribution in the mixture was a function of agitation, and uniformity of distribution on the leaf surface was dependent on mode of the application. Therefore, our test procedure was designed to obtain a satisfactory oil distribution and deposition on leaf surfaces. Dipping leaves in the mixtures and spraying with a hand atomizer resulted in different patterns of deposition (Fig. 3). Dipping gave less uniform deposition than spraying; however, neither method provided a complete coverage of the leaf surface with oil droplets.

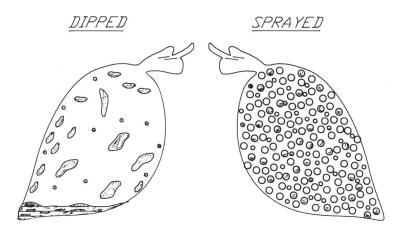

DIPPED   *SPRAYED*

Fig. 3--Distribution of oil droplets on orange leaves.

Test 1

During the 90 min of elasped time it took to evaluate the surface temperature of the dipped leaves, air temperature averaged $32 \pm 2°C$ and varied with passing clouds and wind breezes. Although leaf temperature averaged $3 \pm 1°C$ above air temperature, it was impossible to attribute any variations in leaf temperature to oil type or concentration (data not shown). Very subtle differences in leaf angle may have made relatively large differences in leaf temperatures.

Oil distillation temperature did not have a significant effect on stomatal conductance (Table 2). Averaged over all DTs of oils, there was a significant effect of oil concentration on G_s. Among the oil treatments, both the 4 and 2% had significantly lower G_s than the 1% treatment (0.37, 0.38, and 0.57 cm s^{-1}, respectively). Overall, oil treatment did not affect G_s compared to emulsifier only checks. Within VOLCK Supreme Spray treatment (247°C 50% DT), there was a significant decrease in G_s with increasing concentration. However, the interaction between oil DT and oil concentration was not significant.

TABLE 2 -- The effect of oil distillation temperature and concentration on stomatal conductance of 'Valencia' orange leaves dipped into oil plus emulsifier solutions. Each value is a mean[a], n = 5 (std. dev.).

Oil Type	Oil 50% Distillation Temperature °C (°F)	Oil Plus Emulsifier Concentration %		Stomatal Conductance G_{s-1} cm s
Sunspray 7N	224 (435)	1		0.52a (0.11)
		2		0.40a (0.21)
		4		0.51a (0.33)
			X̄	0.48a (0.22)
Sunspray 9N	235 (455)	1		0.59a (0.27)
		2		0.29a (0.21)
		4		0.38a (0.18)
			X̄	0.42a (0.24)
VOLCK Supreme Spray	247 (476)	1		0.62a (0.36)
		2		0.45ab (0.26)
		4		0.22b (0.09)
			X̄	0.43a (0.29)
Emulsifier T-MULZ-FLO only		1[b]		0.79a (0.52)
		2		0.34a (0.18)
		4		0.46a (0.20)
			X̄	0.53a (0.37)

[a] Means within/among mixtures followed by unlike letters differ significantly at $p < 0.05$, using Duncan's Multiple Range Test.
[b] Has same amounts of emulsifier as in oil-emulsifier solutions.

Test 2

Within the sprayed treatments, there was no significant differences among G_s nor A regardless of treatment (Table 3). Although there may have been a trend of oil sprays reducing net gas exchange below that of the emulsifier only treatment, large leaf to leaf variations obscured any significant differences. Spraying leaves with Sunspray 7N-1% (224°C 50% DT) raised dark respiration (R) rates above those of the control leaves but Sunspray 9N-4% (235°C 50% DT) treatment had lower rates. Thus, there were no trends in the effects of spray oil distillation temperatures or concentrations on dark respiration.

Previous report of reductions in net gas exchange [9] may have been due to spraying treatments of different types of oil. We observed no phytotoxic responses from any treatment after 3 months under glasshouse conditions.

Field Test

In the field test, problems associated with identifying specific sources of leaves on the ground and differences in the size and age of the trees, made it impossible to attribute leaf drop data to spray oil treatments. Overall, there was no obvious phytotoxicity among any of the oil treated trees compared to unsprayed checks.

TABLE 3 -- The effect of oil distillation temperature and concentration on stomatal conductance, net CO_2 assimilation, and dark respiration rates of 'Valencia' orange leaves sprayed with oil plus emulsifier solutions. Each value is a mean[a], n = 5 (std. dev.).

Oil Type	Oil 50% Distillation Temperature $^\circ$C ($^\circ$F)	Oil Plus Emulsifier Concentration %	Stomatal Conductance G_s cm s^{-1}	Net CO_2 Assimil'n A μmol m^{-2}s^{-1}	Dark Respiration R μmol m^{-2}s^{-1}
Sunspray 7N	224 (435)	1	0.24a (0.16)	4.84a (3.3)	0.63a (0.61)
		4	0.16a (0.14)	3.41a (1.9)	0.26ab (.14)
Sunspray 9N	235 (455)	1	0.28a (0.16)	6.56a (3.5)	0.29ab (.14)
		4	0.26a (0.20)	5.41a (3.0)	0.12b (0.26)
VOLCK Supreme Spray	247 (476)	1	0.21a (0.16)	5.16a (3.1)	0.26ab (.12)
		4	0.22a (0.15)	4.63a (2.6)	0.28ab (.13)
Emulsifier only		1[b]	0.34a (0.20)	6.57a (3.1)	0.45ab (.40)

[a]Means followed by unlike letters differ significantly at $p < 0.05$, using Duncan's Multiple Range Test.
[b]Has same amount of emulsifier as in oil-emulsifier solutions.

ACKNOWLEDGMENTS

Florida Agricultural Experiment Station Journal Series No. R-00167.

The authors wish to thank Joseph W. Serdynski and Martin L. Smith, Jr. for their technical assistance.

Trade and company names mentioned in this paper are solely for providing specific information. Their mention does not constitute an endorsement over other products not mentioned.

REFERENCES

[1] Trammel, K. and Simanton, W. A., "Properties of Spray Oils in Relation to Citrus Pest Control in Florida," Florida State Horticultural Society Proceedings, Vol. 79, 1966, pp 12-18.

[2] Jeppson, L. R. and Carman, G. E., "Low Volume Applications to Citrus Trees: Effectiveness in Control of Citrus Red Mite and California Red Scale with Petroleum Oils and Pesticides," Journal of Economic Entomology, Vol. 67, No. 3, 1974, pp 403-407.

[3] Whiteside, J. O., "Action of Oil in the Control of Citrus Greasy Spot," Phytopathology, Vol. 63, No. 2, 1973, pp 262-266.

[4] Whiteside, J. O., "Timing of Single-Spray Treatments for Optimal Control of Greasy Spot on Grapefruit Leaves and Fruit," Plant Disease, Vol. 66, No. 8, 1982, pp 687-689.

[5] Simanton, W. A. and Trammel, K., "Recommended Specifications for Citrus Spray Oils in Florida," Florida State Horticultural Society Proceedings, Vol. 79, 1966, pp 26-30.

[6] Dean, H. A., Tannahill, H., and Bush, J. R., "Effects of Postbloom and Summer Oils on Yield, Fruit Size, and Growth of Four Varieties of Oranges, 1970-74," Journal of Economic Entomology, Vol. 71, No. 2, 1978, pp 211-216.

[7] Krenek, M. R. and King, D. N., "The Relative Phytotoxicity of Selected Hydrocarbon and Oxygenated Solvents and Oils," Pesticide Formulation and Application Systems: 6th Volume, ASTM STP 943, D. I. B. Vander Hooven and L. D. Spicer, Eds., American Society for Testing and Materials, Philadelphia, 1987, pp 3-19.

[8] Wright, N. E. H., "Personal Communications on Horticultural Spray Oils," Sun Refining and Marketing Company, 1988.

[9] Jones, V. P., Youngman, R. R., and Parrella, M. P., "Effect of Selected Acaricides on Photosynthetic Rates of Lemon and Orange Leaves in California," Journal of Economic Entomology, Vol. 76, 1983, pp 1178-1180.

[10] Ferree, D. C., "Influence of Pesticides on Photosynthesis of Crop Plants," in: Photosynthesis and Plant Development, The Hague, 1979, pp 331-341.

[11] Anonymous, "General Considerations in Preparing Emulsions," Sun Refining and Marketing Company.

[12] Syvertsen, J. P., "Light Acclimation in Citrus Leaves. II. CO Assimilation and Light, Water, and Nitrogen Use Efficiency," Journal of the American Society for Horticultural Science, Vol. 109, No. 6, 1984, pp 812-817.

[13] Syvertsen, J. P. and Smith, M. L., Jr., "An Inexpensive Leaf Chamber for Measuring Net Gas Exchange," HortScience, Vol. 18, No. 5, 1983, pp 700-701.

[14] Jarvis, P. G., "The Estimation of Resistance to Carbon Dioxide Transfer," in: Plant Photosynthetic Production: Manual of Methods, The Hague, 1971, pp 566-631.

Alam Sundaram

INFLUENCE OF TWO POLYMERIC ADJUVANTS ON PHYSICAL PROPERTIES, DROPLET
SPREADING AND DRYING RATES, AND FOLIAR UPTAKE AND TRANSLOCATION OF
GLYPHOSATE IN VISION® FORMULATION

REFERENCE: Sundaram, A., "Influence of Two Polymeric Adjuvants
on Physical Properties, Droplet Spreading and Drying Rates,
and Foliar Uptake and Translocation of Glyphosate in Vision®
Formulation", Pesticide Formulations and Application Systems:
10th Volume, ASTM STP 1078, L.E. Bode, J.L. Hazen and D.G.
Chasin, Eds., American Society for Testing and Materials,
Philadelphia, 1990.

ABSTRACT: The effect of Sta-Put® and Silwet® L-7607 on physical
properties (viscosity, surface tension and volatility), droplet
spreading and drying rates, and foliar uptake and translocation of
glyphosate was studied using white birch seedlings and branch
tips. Three end-use mixtures were prepared using Vision®, one
in water alone and the other two with 0.05% of the adjuvants
in water. Physical properties were measured to examine their
roles on droplet spreading and drying rates. Foliar uptake was
investigated to study the effect of droplet spreading and drying rates on foliar retention; and translocation was studied
to examine the role of the two polymers on bioavailability of
glyphosate.

The adjuvants did not contribute to marked differences in
the viscosities or volatilities of the three end-use mixtures,
although the surface tensions were altered to some extent. Silwet caused a marked increase in the droplet spread areas, along
with a simultaneous decrease in the droplet drying time. Sta-
Put did not alter any of these parameters. No simple relationship could be found between surface tensions of the mixtures
and droplet spreading or drying rates. Both adjuvants contributed to an increase in the foliar uptake of glyphosate, but Silwet caused a much greater increase than Sta-Put, thus suggest-
ing a relationship between droplet spreading and foliar uptake.
Nevertheless, there was no significant difference in the amount
of glyphosate translocated between the three end-use mixtures,
thus indicating no evidence of reduced bioavailability because
of the presence of the two polymers at the concentration levels
used in the study.

Dr. Alam Sundaram is a Research Scientist and Project Leader for
Pesticide Formulations Project, at Forestry Canada, Forest Pest Manage-
ment Institute, 1219 Queen Street East, P.O. Box 490, Sault Ste. Marie,
Ontario, Canada P6A 5M7.

KEYWORDS: Glyphosate, herbicide end-use mixtures, polymeric
adjuvants, physical properties, droplet spreading, droplet
drying, foliar uptake, herbicide translocation

Polymers are used in herbicide formulations to provide beneficial
effects, such as, alteration of droplet size spectra for increasing
deposits on target sites [1-3], and for reducing off-target drift [4-8].
However, certain polymers can provide adverse side effects, viz., reduce herbicidal activity via entrapment of the herbicide molecules into
the polymeric structure [9]. Polymers have also increased droplet spreading on foliage, provided a greater area of contact and enhanced the
rate of uptake [10]. On the other hand, polymers can also accelerate
droplet drying rates on foliage [11] and reduce uptake, since the loss
of a liquid phase on foliage has been known to reduce cuticular penetration [12].

Sta-Put® is a polymeric adjuvant (Nalco Chemical Co., Illinois,
USA) that can alter the droplet size spectra of the spray cloud indicating its potential as a "drift retardant" adjuvant [13] for herbicide
tank mixes. Silwet® L-7607 is a non-ionic organosilicone copolymer surfactant (Union Carbide, Montreal, Canada) which can increase droplet
spreading on foliage by reducing the contact angle [10,11]. However,
data are sparse in the literature on the ability of these polymers to
reduce the bioavailability of herbicides, and to alter the physical
properties (viscosity, surface tension and volatility) [14] of the end-
use mixes, thereby affecting droplet spreading/drying rates, foliar
uptake and translocation.

The purpose of the present study was to investigate the effect of
Sta-Put and Silwet polymers on physical properties, droplet spreading
and drying rates, foliar uptake and translocation of glyphosate [N-
(phosphonomethyl)glycine, Monsanto Agricultural Products Company,
St. Louis, Missouri, USA] in the end-use mixtures of Vision® formulation [a formulation concentrate of glyphosate containing 356 g of active
ingredient (AI) per liter], using white birch, *Betula papyrifera* (Marsh.)
seedlings and branch tips.

EXPERIMENTAL METHODS

End-use Mixtures

Glyphosate end-use mixtures and other materials used in the study
are listed in Table 1, along with the percentage compositions of the
ingredients used in preparing them.

Physical Properties

Physical properties measured were viscosity, surface tension and
volatility. Viscosity was measured relative to water (relative viscosity) using Ostwald viscometer [15]. Surface tension and volatility
were measured as described by Sundaram and Leung [16]. Volatility parameters are expressed in rate of evaporation, R(Evap), half-life (i.e.,
$T_{\frac{1}{2}}$, the time required for the volatile components to reach 50% of their
initial concentrations), and the percent non-volatile componets (NVC%,

Table 1 - Percentage compositions of ingredients used in the glyphosate end-use mixtures

Formulation ingredients	Percentage composition (v/v)	
	Mixture VW	Mixture VWSt-0.05 [a] or VWSi-0.05 [a]
Distilled water:	54.19	54.14
Vision® formulation:	14.04 [b]	14.04
14C-glyphosate: [c]	31.77	31.77
Adjuvant:	---	0.05

a: The formulation VWSt-0.05 contained the Sta-Put adjuvant whereas the VWSi-0.05 formulation contained the Silwet L-7607 adjuvant.

b: At a concentration level of 14.04 ml of Vision in 100 ml of the end-use mixture the dosage rate is equivalent to 0.5 kg of active ingredient in 10 liter per ha.

c: The radiolabeled product had a specific activity of 49 µCi per mg of glyphosate in 13.7 ml of solution. It contained the same iso-propylamine salt of glyphosate as in Vision, along with the same surfactant in the same weight ratio.

the residual amounts which were left unevaporated until at least 120 h). All measurements were carried out at room temperature (20 ± 2°C). The data are given in Table 2.

Droplet Spreading and Drying Rates

To investigate droplet spreading and drying rates, aliquots of the three mixtures, VW, VWSt-0.05 and VWSi-0.05 (see Table 1 for the description of these mixtures), were mixed with a water-soluble fluorescent dye (Erio Acid Red, St. Lawrence Aniline Company, Brockville, Ont., Canada), at 0.2 g per 100 ml. This was intended to facilitate ready visualization of the droplet under a microscope on different surfaces. The addition of this minute quantity of the dye was tested in a previous experiment, and was found not to alter the physical properties of the three mixtures under investigation.

a. Equilibrium Spread Areas and the Associated Rates

White birch seedlings (6 month old, height 44 ± 7 cm or 66 ± 7 cm with the pot), raised from seed under greenhouse conditions, were selected for the study. The average number of leaves per seedling was 14 ± 3. Prior to the start of the study, the seedlings were allowed to acclimatize for one week in the treatment area, a part of the greenhouse maintained at a temperature of 20 ± 2°C, relative humidity of 75 ± 5%, and a photoperiod of 16 h light and 8 h darkness. The average surface area

Table 2 - Physical properties of the three end-use mixtures used in the study at 20 ± 2°C and 75 ± 5% relative humidity

Physical properties	End-use mixtures used		
	VW	VWSt-0.05	VWSi-0.05
Relative viscosity	1.38	1.68	3.63
Surface tension (mN/m)	31.5	37.2	27.5
Volatility data:			
i. R(Evap) [a]	3.95	3.23	3.09
ii. T½ (min) [b]	11.0	13.1	13.4
iii. NVC% [c]	13.0	15.3	16.9

a: Percentage weight decrease of the liquid film per min, as calculated during the initial 10 min of evaporation.

b: Half-life, T½, refers to the time required for the volatile components of the mixtures to evaporate down to 50% of their initial values.

c: The residual amounts which were left unevaporated until at least 120 h after the start of the experiment.

of the leaves at the mid-crown level was 100 ± 5 cm^2 per leaf at the time of the study.

Droplets of 0.5 μl in volume (or 1000 μm in diameter) were produced using a precision micro-applicator (Instrumentation Specialties Company, 4700 Superior Lincoln, NB 68504, USA). These were collected on the adaxial surface of a mid-crown leaf at the rate of 8 droplets per 6 cm^2. The droplets were observed under a microscope using dim illumination to avoid undue heating of the droplet (which would increase the rate of evaporation); and the degree of spreading was examined. When the diameter of the spread area attained a constant value (the equilibrium state of spreading), the diameter was measured, and the time to reach the equilibrium state was noted. These procedures were replicated several times (minimum number of leaves used for each mixture was 10, equivalent to 80 droplets; maximum was 20 or 160 droplets), and the mean ± SD values of the spread areas were calculated for the three mixtures described in Table 1. In addition, two artificial surfaces, glass plate and an acetate transparency sheet (each 2.5 cm x 2.5 cm) (the transparency sheet was obtained from General Photography Ltd., 1350 Birchmont Road, Scarborough, Ontario, Canada) were used for the sake of comparison. The spread factor data (SF) were calculated using the equation:

$$SF = \frac{\text{Diameter of the spread area on a surface}}{\text{Diameter of the spherical droplet that would be airborne}}$$

The data on SF values and the time required for equilibrium spreading are given in Table 3.

Table 3 - Spread factor data for the three end-use mixtures on different surfaces and the time required for complete spreading of droplet[a] - temperature 20 ± 2°C and rel. humidity 75 ± 5%

Surface type	End-use mixtures used		
	VW	VWSt-0.05	VWSi-0.05
Spread factor values[b]			
1. Birch leaf	2.03 (± 0.13)	2.02 (± 0.11)	4.55 (± 0.13)
Relative spread area (RSA) [c]	1	1	4.93
2. Glass plate	2.63 (± 0.11)	2.42 (± 0.09)	5.20 (± 0.14)
Relative spread area (RSA) [c]	1	1	4.33
3. Acetate sheet	1.54 (± 0.04)	1.54 (± 0.03)	3.60 (± 0.04)
Relative spread area (RSA) [c]	1	1	5.46
Time (min) required for complete spreading [b]			
1. Birch leaf	15.3 (± 0.60)	16.9 (± 0.70)	8.00 (± 0.65)
2. Glass plate	7.03 (± 0.45)	9.07 (± 0.53)	4.59 (± 0.37)
3. Acetate sheet	10.2 (± 0.45)	12.0 (± 0.53)	7.05 (± 0.37)

a: The size of the droplet used was 0.5 µl in volume or 1000 µm in diameter.

b: The data represent the mean ± SD of at least 80 replicate droplet measurements.

c: Calculated as:

$$RSA = \frac{\pi/4\ (\text{diam. of spread area for VW with adjuvant})^2}{\pi/4\ (\text{diam. of spread area for VW without adjuvant})^2}$$

In some instances, the spread area of the droplet failed to form a circular stain on the foliar surface. In these cases, the total area of the stain was determined microscopically, and the diameter of the circle that would have the same area as the non-circular stain on foliage was calculated for computing the spread factor.

b. Droplet Drying Rates

For droplet drying rates, a method was developed based on the principle that liquids can be sheared by application of pressure much more easily than solids. If a liquid droplet was collected on a leaf surface the droplet would spread, evaporate and penetrate into leaf cuticle at the same time. If the initial ingredients of the droplet were non-volatile and volatile liquids, the evaporated droplet would leave a liquid residue of diameter D1 on the leaf surface. Upon application of pressure, the diameter of the droplet would noticeably increase to D2. In contrast, if the initial ingredients were solids and volatile liquids, the evaporated droplet would leave a solid residue (with a diameter D1), but when pressure was applied the diameter would not increase noticeably. These principles were used to develop a method for determining the time required for complete drying of droplets on surfaces.

Droplets of 0.5 μl in volume were generated as described above, and were collected on a 6-cm^2 area of a mid-crown birch leaf. The drying process was observed with respect to time, visually during the initial stages, i.e., until the shiny appearance (due to reflection of light) of the droplet surface had lasted. The droplets were then left to dry for another 2 min, and the diameter (D1) of the spread area was measured microscopically. The time lapse (T1), i.e., from the time of droplet collection on the leaf until diameter measurement, was noted. Since this stage may not represent a completely evaporated droplet, further testing for droplet dryness was carried out by placing a clean acetate sheet (6 cm^2 area) over the birch leaf. A force of 100 g-weight was applied gently onto the top acetate sheet without causing any lateral movement of the assembly. After 2 min, the weight was removed and the assembly was placed under a microscope for diameter (D2) measurement of the droplet spread area. If D2 was the same as D1, then the droplets were completely dry and consisted of a solid phase, and the time T1 was taken as that required for complete drying of the droplet. If D2 was greater than D1, this indicated a liquid droplet (either a partially evaporated droplet or a droplet consisting of a non-volatile liquid phase). Therefore, the experiment was repeated using a fresh birch leaf. This time, the drying process was observed for a period (T2) which was 15 min longer than T1 (i.e., T2 = T1 + 15), and the entire process was carried out. After a lapse of T2 min, the diameter D3 was measured and another acetate sheet was placed, followed by the 100 g-weight. Two min later the diameter D4 was measured. If D4 was equal to D3, then the droplets were dry, and the time T2 was taken as required for complete drying of droplets. This procedure was repeated 'n' times until Dn became equal to D_{n-1}. The time that was needed to attain the constant D_{n-1} value was taken as that required for the complete (or maximal) drying of the droplet. Following this procedure, the droplet drying times were determined for the three end-use mixtures listed in Table 1, and the data are given in Table 4.

In addition to the birch leaf, glass plate and acetate sheet surfaces were also used for the investigation in the same manner (Table 4).

Table 4 - Time required for droplet^a drying on birch leaf, glass plate
and acetate sheet for the three end-use mixtures used in the
study - temperature 20°C, rel. humidity 75 ± 5%

Surface type	End-use mixtures studied		
	VW	VWSt-0.05	VWSi-0.05
Birch leaf	> 360^b	> 360	240
Glass plate	> 360	> 360	180
Acetate sheet	> 360	> 360	180

a: The size of the spherical droplet was 1000 μm in diameter.

b: The data represent the mean of at least 10 replicates.

Foliar Uptake and Translocation

Foliar uptake and translocation were investigated using two types
of studies, viz., using 6-month old birch seedlings and branch tips
cut out from the 6-month old seedlings.

a. Study I - Using Birch Seedlings

Twenty-six seedlings (description same as that under "Equilibrium
Spread Areas and the Associated Rates") were selected for Study I. The
seedlings were divided into four groups, A, B, C and D. Groups A to C
consisted of 8 seedlings each, and group D, of 2 seedlings. Group A received the VW mixture; group B, VWSt-0.05; and group C, VWSi-0.05. Group
D served as the control, for measuring the background radioactivity in
the plants.

A 40 μl volume (containing 101000 disintegration per min [dpm] of
radioactivity) of the three end-use mixtures (Table 1) was applied to
each seedling in 80 x 0.5 μl droplets (each 1000 μm in diameter) to the
adaxial surface of four mid-crown leaves, at the rate of 20 droplets
per leaf (or 0.2 droplet per cm^2), using the micro-applicator described
earlier. This dosage regime provided an application rate of 0.5 kg AI
in 10 liter per ha area of the treated leaves.

Out of the eight seedlings used for each formulation, two were harvested at each of the four post-treatment periods, i.e., 8, 24, 48 and
72 h respectively. Each plant was divided into four segments, viz.,
leaves above treated leaves, stem, leaves below treated leaves and root.
The adaxial surface of the treated leaves were individually rinsed with
2 x 20 ml of distilled water into a graduated cylinder (the time duration for each rinsing was 30 s). This procedure was repeated for the

other three treated leaves and the total rinsings were pooled. Exactly 3 ml volume of the rinse solution (described as the treated-leaf wash) was transferred into a scintillation vial containing 17 ml of a liquid scintillant (Scinti Verse[TM] II, SO-X-12, Fisher Scientific Company, New Jersey, USA) for ^{14}C-assay. All plant parts [including the washed treated leaves (described as the treated-leaf residue)] were then oven-dried for 4 h at 60°C, weighed and combusted in a biological oxidizer (Packard Oxidizer, Model 306, United Technologies Packard, Packard Instrument Company, Illinois, USA). The evolved [14]CO$_2$ was absorbed in vials containing Carbosorb® (an aqueous scintillant for absorbing [14]CO$_2$, United Technologies Packard, Illinois, USA) for ^{14}C-assay.

The soil in the pot was not assayed for radioactivity because in a previous study soil samples showed little ^{14}C-content, indicating neg-ligible exudation of glyphosate via root (unpublished data). The radio-activity of all plant samples and leaf extracts was determined by a Beckman LS9000 liquid scintillation counter with a built-in automatic external standardization to determine counting efficiency. The range of counting efficiency was 93 to 97%, and the data in Table 5 were correc-ted for these factors. Because few glyphosate metabolites have been reported in plants within 72 h after treatment [17,18] the radioactivity recovered will be referred to as ^{14}C-glyphosate.

b. Study II - Using Birch Branch Tips

Because only two seedlings were used at each sampling time for each end-use mixture, only a preliminary estimate of ^{14}C-distribution in plant sections can be made at different time intervals (Table 5). To obtain more meaningful information on differences in foliar uptake and bioavailability between the three end-use mixtures, several seed-lings would be required per mixture. However, the use of a large number of seedlings would involve extensive labour, time and cost of the mate-rials. To overcome this problem, small branch tips were used in Study II.

Twenty-six branch tips (each 20 cm long containing three fully deve-loped young leaves) were clipped from the top portion of the 6-month old seedlings (one branch tip from each seedling). The under-developed young leaves, except the shoots, were removed and discarded, leaving only the three fully-developed leaves and shoots in the branch tip. The stem of each branch was placed at once in a 50 ml capacity plastic vial contai-ning tap water, and the branch was supported upright by a tubing and a lid with a hole. Similar branch clippings were tested for their survi-val rate and growth patterns for a period up to 72 h in a preliminary investigation prior to the actual study; and it was observed that the branches remained quite healthy and showed small but significant amount of weight gain during this period.

Twenty-four branches were equally divided into 3 groups, G, H and J. Group G received VW mixture, group H, VWSt-0.05 and group J, VWSi-0.05. The remaining two branches (group K) served as the control for measuring the background radioactivity in the plants. A 10-μl volume (containing 25250 dpm of radioactivity) of the mixture was applied to one leaf (sur-face area 100 cm^2). For relevant details see under 'Study I'.

The eight branches used for each formulation were harvested at 48 h post-treatment. This time period was considered as adequate for detecting

Table 5. Foliar uptake and translocation of glyphosate into different parts of white birch seedlings following treatment with three end-use mixtures

Sample description	End-use mixtures	Percentage distribution*			
		8h	24h	48h	72h
Treated leaves	VW	24.6 ± 5.0	38.6 ± 6.5	42.4 ± 1.4	43.8 ± 3.8
	VWSt-0.05	35.7 ± 0.2	44.5 ± 6.5	46.2 ± 3.3	57.1 ± 15.2
	VWSi-0.05	38.9 ± 1.3	49.7 ± 3.5	52.5 ± 4.4	63.5 ± 10.2
Leaves above treated leaves	VW	0.1 ± 0.02	0.5 ± 0.02	1.1 ± 0.7	0.9 ± 0.1
	VWSt-0.05	0.2 ± 0.01	0.4 ± 0.10	0.6 ± 0.4	0.8 ± 0.6
	VWSi-0.05	0.2 ± 0.05	0.5 ± 0.10	0.7 ± 0.3	1.0 ± 0.6
Leaves below treated leaves	VW	0.1 ± 0.03	0.2 ± 0.01	0.2 ± 0.02	0.4 ± 0.20
	VWSt-0.05	0.3 ± 0.20	0.3 ± 0.10	0.2 ± 0.03	0.2 ± 0.01
	VWSi-0.05	0.3 ± 0.10	0.3 ± 0.10	0.3 ± 0.05	0.3 ± 0.10
Stem	VW	2.9 ± 0.9	7.6 ± 1.8	8.9 ± 3.1	12.1 ± 0.6
	VWSt-0.05	2.5 ± 0.5	7.3 ± 1.1	9.1 ± 2.8	8.5 ± 1.0
	VWSi-0.05	2.8 ± 0.6	7.1 ± 1.0	8.9 ± 1.9	10.5 ± 0.9
Root	VW	2.4 ± 0.4	4.4 ± 2.6	5.6 ± 0.7	8.0 ± 2.0
	VWSt-0.05	2.8 ± 1.3	6.7 ± 2.6	10.5 ± 0.8	11.2 ± 3.0
	VWSi-0.05	3.2 ± 1.2	7.7 ± 1.5	11.5 ± 0.9	13.9 ± 3.3
Plant total	VW	30.1	51.3	58.2	65.2
	VWSt-0.05	41.5	59.3	66.6	77.8
	VWSi-0.05	45.4	65.3	73.9	89.2
Leaf wash	VW	69.9 ± 3.8	48.7 ± 10.8	41.8 ± 4.4	34.8 ± 2.5
	VWSt-0.05	58.5 ± 1.7	40.7 ± 5.7	33.4 ± 7.3	22.2 ± 6.8
	VWSi-0.05	54.6 ± 2.0	34.7 ± 4.7	26.1 ± 5.7	10.8 ± 3.3

* Values represent the mean of two sets of data obtained from the trees.
All values were corrected for the ^{14}C-counting efficiency.
Percentage distribution values were calculated as:

$$\% \text{ Distribution} = \frac{\text{Radioactivity recovered in sample}}{\text{Total radioactivity recovered}} \times 100$$

differences in the uptake and translocation patterns between the three
formulations used. Each branch tip was divided into two parts, viz.,
treated leaf and the remaining parts. The tap water in the vial was
also collected for radioassay to examine glyphosate movement via stem
into water. The treated leaf was rinsed as described above to provide
treated-leaf residue and treated-leaf wash. All samples were assayed
for ^{14}C-glyphosate in the same manner as mentioned in 'Study I'. The
data are given in Table 6.

RESULTS AND DISCUSSION

Influence of Viscosity, Surface Tension and Volatility Parameters on
Droplet Spreading and the Time Required for Complete Spreading

The data in Table 2 indicate that viscosities of the two end-use
mixtures, VW and VWSt-0.05, were similar to each other, and that that
of VWSi-0.05 was only slightly higher. Such small differences are not
expected to contribute to marked differences in droplet spreading [14].
The surface tension of VWSt-0.05 was higher than that of VW. This could
be due to interactions [19] between the Sta-Put polymer and the surfactant present in Vision. However, the equilibrium spread areas were similar for VW and VWSt-0.05 on all three surfaces tested. In contrast, the
surface tension of VWSi-0.05 was lower than those of the other two mixtures, and the droplet spread areas were markedly greater (ca. 4 to 5
times) than those of VW or VWSt-0.05. The data thus indicate the lack of
a simple relationship between surface tensions and droplet spread areas,
although the lowest surface tension of VWSi-0.05 could have contributed
to the highest degree of droplet spreading on all the three surfaces.
The chemical nature of the adjuvants appear to have played a more important role on droplet spreading than the physical properties. However,
further investigations using different concentrations of the same adjuvant are necessary before any definite conclusions can be drawn concerning the role of surface tension on droplet spreading. The volatilities
of the three mixtures, similarly to the viscosities, were comparable to
one another and therefore, would not contribute to marked differences
in the droplet spreading characteristics of the three end-use mixtures.

The present findings are in agreement with those of Zabkiewicz et
al. [10,11] and of Sands and Bachelard [20] who found that the chemical
nature of adjuvants play an important role on droplet contact angle,
wettability and equilibrium spread areas, although none of these authors
determined the three physical properties measured in the present study.

The data on droplet spreading (Table 3) indicate that the influence
of the surface tension was not as pronounced as the differences in the
surface characteristics of the three collectors. For example, the spread
factor data on glass plate were significantly higher (ANOVA $P \leq 0.05$)
than the values on the acetate sheet. Similarly, slight but significant
differences were observed in the spread factor data on birch leaf and
glass plate. Thus, out of the three surfaces tested, the acetate sheet
provided the lowest spread factor values for all three mixtures.

The data on the time required for complete spreading of the droplet
(Table 3) showed significantly lower (ANOVA $P \leq 0.05$) values for VWSi-
0.05 than for VW or VWSt-0.05, on all three surfaces. The reason for

Table 6. Foliar uptake and translocation of glyphosate into birch branch tips at 48 h after treatment with three end-use mixtures

Sample description	Percentage distribution of radioactivity* for mixtures		
	VW	VWSt-0.05	VWSi-0.05
Treated leaf	39.9 ± 2.62 ^{a}	44.1 ± 2.88 ^{b}	52.1 ± 2.94 ^{c}
Remaining parts	1.31 ± 0.42 ^{d}	1.65 ± 0.58 ^{dg}	2.01 ± 0.2 ^{g}
Leaf wash	58.8 ± 2.66 ^{h}	54.2 ± 2.45 ^{j}	45.9 ± 2.86 ^{k}
Tap water in vial	---	---	---

* Values represent the mean ± SD obtained from eight sets of data obtained from eight branch tips used for each end-use mixture.
All values were corrected for the ^{14}C-counting efficiency.
Percentage distribution values were calculated as:

$$\% \text{ Distribution} = \frac{\text{Radioactivity recovered in sample}}{\text{Total radioactivity recovered}} \times 100$$

[a-k] Values with the same superscript letters are not significantly different from one another (Student-Newnam-Keuls test, error rate $\propto > 0.05$).

this lies in the very high values of droplet spread areas observed for
VWSi-0.05. The higher the droplet spreading, the faster the evaporation of volatile components, and therefore, less amount of time would
be required for complete spreading of the droplet. In summary, the present data emphasize the complexity of the interactions between surface
tension, volatility and the physicochemical nature of the collection
surfaces on the spreading capability of the different liquids.

Influence of Physical Properties of the End-use Mixtures on Droplet Drying Times

The data in Table 4 indicate markedly lower drying times for the
droplets of VWSi-0.05 than for VW or VWSt-0.05. Regardless of the nature of the surfaces used, VWSi-0.05 dried the fastest, i.e., in about
3 to 4 h after treatment, whereas, VW and VWSt-0.05 did not show complete drying even after 6 h. The reason for this behavior was the formation of a thick waxy crust over the droplets of VW or VWSt-0.05 (this
layer was visible when viewed under a microscope). This layer must have
slowed down the rate of evaporation of water molecules escaping into
the ambient air. Consequently, the inner core of the droplet remained
as liquid, and the droplets failed to dry completely even after 6 h.
Such a crusty layer was not visible in the case of VWSi-0.05, probably
because of excessive spreading, and that is perhaps why the droplets
were dried out completely in 3 to 4 h after treatment.

Foliar Uptake and Translocation of Glyphosate

a. Study I - Using Birch Seedlings

The data in Table 5 indicate that foliar uptake of glyphosate is
relatively a slow process since more than 50% was washed off into the
leaf rinse at 8 h after treatment, irrespective of the end-use mixture
tested. However, as the exposure duration increased, the uptake gradually increased to 44 to 64% depending on the adjuvant, and only about
10 to 35% was washed off at 72 h.

Similar to the foliar uptake, the translocation of glyphosate from
the treated leaf into other parts of the plants is also shown to be
slow because only about 6% of the applied amount was translocated at 8 h
after treatment, and about 94% remained in the treated leaf. However,
with the increased exposure, translocation increased gradually and reached 21 to 25% at 72 h (Table 5). Nevertheless, the treated leaf still
contained about 75 to 79% of the applied amount, thus indicating incom-
plete translocation even after 72 h. The amount of radioactivity detected in the stem and root sections increased gradually from the 8 h value to about 4 to 5 times higher at 72 h. The present findings are in
agreement with those reported in the literature [21], although the
amount absorbed and translocated was higher in the present study than
in those reported.

Regarding the influence of the two polymeric adjuvants on the bioavailability of glyphosate, the present data indicate no evidence of
glyphosate entrapment into the polymeric chain, thereby making it less

bioavailable. On the contrary, the two adjuvants contributed to a significant increase in foliar uptake, as indicated by analysis of variance test (ANOVA P ≤ 0.05), but yet the translocated amount seemed to be similar for all three end-use mixtures (Table 5).

b. Study II - Using Birch Branch Tips

The results obtained from the eight branch tips used in Study II for each mixture were subjected to statistical treatment using the Studen-Newman Keuls test (S-N-K) [22]. The data indicated that on average, ca. 60% of the applied amount was washed off at 48 h from the leaf treated with VW, as opposed to 54% with VWSt-0.05 and 46% with VWSi-0.05 (Table 6), thus indicating a significant increase (S-N-K test, error rate α ≤ 0.05) in the 'apparent foliar uptake' because of the presence of the two adjuvants. The amount translocated into the remaining parts of the branch tip was low and ranged from 1.3% (for VW) to 2.0% (for VWSi-0.05). No significant difference was noted in the translocated amount between VW and VWSt-0.05, or between VWSt-0.05 and VWSi-0.05; but a significantly higher amount was translocated due to the presence of the Silwet adjuvant in VWSi-0.05, compared to the VW mixture.

Little radioactivity was detected in the tap water in vial, thus indicating negligible movement of glyphosate _via_ stem into water.

The significant increase in the 'apparent foliar uptake' of glyphosate arising from the presence of the Sta-Put and Silwet polymers, does not necessarily indicate an increase in penetration of glyphosate through the leaf cuticle, since the adjuvants could have simply provided a protective layer over the droplets, thus reducing the amount being washed off during rinsing. Without detailed investigations using extracted plant cuticle [23], it would not be possible to conclude whether the two polymers actually increased the foliar uptake of glyphosate, or simply provided a protective film over the droplets. Nevertheless, the present study indicated no evidence of reduced bioavailability of glyphosate _via_ adsorption or entrapment into the polymeric structure of Sta-Put or Silwet L-7607 at the concentration levels used in the study.

Influence of Droplet Spreading and Drying on Foliar Uptake and Translocation of Glyphosate

The droplet spreading patterns and drying times were very similar for VW and VWSt-0.05, indicating that the Sta-Put adjuvant did not contribute to any changes in these parameters. However, the adjuvants caused a significant increase (by ca. 4%) in the apparent foliar uptake. In contrast, Silwet L-7607 contributed to a marked increase in droplet spreading compared to Sta-Put, and therefore is expected to enhance foliar uptake markedly. The observed data are in agreement with this, since the 'apparent foliar uptake' was 8% higher for VWSi-0.05 than for VWSt-0.05, and 12% higher than for VW alone (Table 6). The present study suggests the possible advantages of adding Silwet L-7607 adjuvant to herbicide mixtures, because it enhanced droplet spreading on target foliage, decreased the amount of glyphosate being wa ed off, and increased foliar retention, although the chemical nature of Silwet L-7607 could have contributed to the higher foliar uptake. Without detailed investigations on foliar uptake using several Silwet® adjuvants which would cause different degrees of droplet spreading, it is not possible to conclude that enhanced droplet spreading would likely increase foliar uptake of herbicides.

The fact that the droplet of VWSi-0.05 dried sooner than those of VW or VWSt-0.05 did not seem to affect foliar uptake and translocation (Tables 4 and 6), an observation in contrast with that reported [12]. The reason for this deviation is not clear and requires further investigations.

ACKNOWLEDGEMENTS

The author wishes to acknowledge with thanks the help of Mr. Donald Buckley in providing the white birch seedlings; and Messers J.W. Leung, J.K. Hatherley and J. Studens for their technical assistance in carrying out the numerous measurements required for this investigation.

REFERENCES

[1]   Sundaram, A., Leung, J.W. and Curry, R.D., "Influence of Adjuvants on Physicochemical Properties, Droplet Size Spectra and Deposit Patterns: Relevance in Pesticide Applications", Journal of Environmental Science and Health, Vol. B22, No. 3, May/June 1987, pp. 319-346.

[2]   Sparks, B.D., Sundaram, A., Kotlyar, L., Leung, J.W. and Curry, R.D., "Physicochemical Properties, Atomization and Deposition Patterns of Some Newtonian Spray Mixtures of Glyphosate Containing Two Spray Modifier Adjuvants", Journal of Environmental Science and Health, Vol. B23, No. 3, May/June, 1988, pp. 235-266.

[3]   Goering, C.E. and Butler, B.J., "Paired Field Studies of Herbicide Drift", Transactions of the ASAE, Vol. 18, 1975, pp. 27-34.

[4]   McNulty, J.J., Anderson, G.W. and Stephenson, G.R., "Evaluation of Low Pressure Nozzles and Nalco-Trol for Reducing Drift with 2,4-D", Canadian Weed Council Research Reports, 1977, p. 316.

[5]   Stephenson, G.R., Ellise, P., Desai, P.D. and Curtis, L.R., "Evaluation of Nalco-Trol for Reducing Herbicide Drift in Roadside Spraying", Abstracts, Weed Science Society of America, Vol. 116, 1977, p. 59.

[6]   Yates, W.E., Akesson, N.B. and Bayer, D., "Effects of Spray Adjuvants on Drift Hazards", Transactions of the ASAE, Vol. 19, 1976, pp. 41-46.

[7]   Yates, W.E., Cowden, R.E. and Akesson, N.B., "Effects of Nalco-Trol on Atomization", USDA Forest Service, Forest Pest Management Report, FPM-85-2, Davis, California, 1985, 37 pp.

[8]   Richardson, R.G., "Control of Spray Drift with Thickening Agents", Journal of Agricultural Engineering Research, Vol. 19, pp. 227-231.

[9]   Doub, J.P., Wilson, H.P. and Hatzios, K.K., "Comparative Efficacy of Two Formulations of Alachlor and Metalachlor", Weed Science, Vol. 36, pp. 221-226.

[10] Zabkiewicz, J.A., Gaskin, R.E. and Balneaves, J.M., "Effect of Additives on Foliar Wetting and Uptake of Glyphosate into Gorse (Ulex europaeus), in Application and Biology, BCPC Monograph No. 28, E.S.E. Southcombe, Ed., Proceedings of a Symposium,

British Crop Protection Council, Croydon, England, 1985, pp. 127-134.

[11] Zabkiewicz, J.A., Coupland, D. and Ede, F., "Effects of Surfactants on Droplet Spreading and Drying Rates in Relation to Foliar Uptake", in Pesticide Formulations: Innovations and Development, American Chemical Society Symposium Series, Vol. 371, American Chemical Society Publications, Washington, D.C., 1988, pp. 77-89.

[12] Hess, F.D., "Herbicide Absorption and Translocation and Their Relationship to Plant Tolerance", in Weed Physiology, Vol. II, Herbicide Physiology, S.O. Duke, Ed., CRC Press, Inc., Boca Raton, Florida, 1984, pp. 191-215.

[13] Bouse, L.F., Carlton, J.B. and Jank, P.C., "Use of Polymers for Control of Spray Droplet Size", American Society of Agricultural Engineers, Paper No. AA-86-005, St. Joseph, Michigan, USA, 1986, 18 pp.

[14] Sundaram, A., "Drop Size Spectra, Spreading and Adhesion; and Physical Properties of Eight Bacillus thuringiensis Formulations Following Spray Application Under Laboratory Conditions", in 9th Symposium on Pesticide Formulations and Application Systems "International Aspects", ASTM STP 1036, J.L. Hazen and D.A. Hovde, Eds., American Society for Testing and Materials, Philadelphia, 1989, pp. 129-141.

[15] Glasstone, S., Textbook of Physical Chemistry, 2nd edition, Macmillan and Co., London, England, 1955, p. 498.

[16] Sundaram, A. and Leung, J.W., "A Simple Method to Determine Relative Volatilities of Aqueous Formulations of Pesticides", Journal of Environmental Science and Health, Vol. B21, No. 2, March/April, 1986, pp. 165-190.

[17] Devine, M.D. and Bandeen, J.D., Fate of Glyphosate in Agropyron repens (L.) Beauv. Growing Under Low Temperatures", Weed Research, Vol. 23, 1983, pp. 69-75.

[18] Zandstra, B.H. and Nishimoto, R.K., "Movement and Activity of Glyphosate in Purple Nutsedge", Weed Science, Vol. 25, 1977, pp. 268-274.

[19] Neustadter, E.L. "Surfactants in Enhanced Oil Recovery", in Surfactants, Th. F. Tadros, Ed., Academic Press, Inc., London, England, 1984, pp. 277-285.

[20] Sands, R. and Bachelard, E.P., "Uptake of Picloram by Eucalypt Leaf Discs", New Phytol., Vol. 72, 1973, pp. 69-86.

[21] Masiunas, J.B. and Weller, S.C., "Glyphosate Activity in Potato (Solanum tuberosum) Under Different Temperature Regimes and Light Levels", Weed Science, Vol. 36, 1988, pp. 137-140.

[22] Steel, R.G.D. and Torrie, J.H., Principles and Procedures of Statistics: A Biometrical Approach, 2nd edn., McGraw-Hill Book Company, New York, 1980, pp. 172-194.

[23] Baker, E.A., Hunt, G.M. and Stevens, P.J.G., "Studies of Plant Cuticle and Spray Droplet Interactions: A Fresh Approach", Pesticide Science, Vol. 14, 1983, pp. 645-658.

Application Systems

Kenneth A. Sudduth, John W. Hummel, and Robert C. Funk

SOIL ORGANIC MATTER SENSING FOR PRECISION HERBICIDE APPLICATION

REFERENCE: Sudduth, K. A., Hummel, J. W., and Funk, R. C.,
"Soil Organic Matter Sensing for Precision Herbicide Appli-
cation," Pesticide Formulations and Application Systems:
10th Volume, ASTM STP 1078, L. E. Bode, J. L. Hazen, and D.
G. Chasin, Eds., American Society for Testing and Materials,
Philadelphia, 1990.

ABSTRACT: A prototype sensor which used near infrared (NIR)
reflectance techniques to determine the organic matter content of the surface layer of soil was developed and was
tested both in the laboratory and in the field. Laboratory
predictions yielded an r^2 of 0.89 and a standard error of
prediction of 0.23% organic carbon (0.40% organic matter)
with 30 representative Illinois soils. Limited in-furrow
field operation of the sensor produced a much higher standard error (0.53% organic carbon), due to the movement of
the soil past the sensor as scanning was accomplished.
Although improvements in sample presentation techniques are
required before the sensor can be used in real-time field
operation, the current configuration could provide rapid,
on-site organic matter determination for mapping and subsequent application rate control using a positioning system.

KEYWORDS: organic matter, near infrared, diffuse reflectance, pesticide application, multivariate calibration

     Precision application of agricultural pesticides is becoming
increasingly important. Misapplication can result in unnecessary
chemical costs and yield reductions due to crop damage or uncontrolled
competition by pests. Application of excess pesticide is also a concern due to the possible pollution of groundwater supplies.

K. A. Sudduth is Agricultural Engineer, USDA-Agricultural Research
Service, Agric. Engineering Bldg., Univ. of Missouri, Columbia, MO
65211; J. W. Hummel is Supervisory Agricultural Engineer, USDA-Agricultural Research Service, 376 AESB, Univ. of Illinois, 1304 W. Penn-
sylvania Ave., Urbana, IL 61801; R. C. Funk is President, AGMED Engineering Services, 3420 Constitution Dr., Springfield, IL 62707.

Rates for soil-applied herbicides must often be increased on soils with higher organic matter contents because the cation exchange capacity of the organic matter increases the adsorption of the herbicides, and thus decreases their phytotoxicity [1]. In one experiment a linear increase in herbicide dose was required to maintain weed control as organic matter increased up to a maximum of 15%, when using compounds such as *s*-triazines, thiocarbamates, and pyridazinones [2].

A common herbicide application practice is adjustment of the application rate to the maximum prescribed for the range of organic matter contents present in a given field or on a particular farm, causing over-application on those areas with lower organic matter contents. This misapplication problem and the associated economic and environmental disadvantages could be avoided if soil organic matter content were sensed and the application rate were then adjusted based on this organic matter signal and the label directions for the particular herbicide. Organic matter sensing might also provide an opportunity to use formulations which have efficacy characteristics highly dependent on maintaining the proper relationship between application rate and organic matter.

Higher organic matter content soils are generally less reflective than soils with lower organic matter contents [1,3,4]. Over the last decade, researchers have developed optical organic matter sensors, based on this reflectance difference, with varying degrees of success.

Krishnan [5] correlated reflectance characteristics in the 400 to 2400 nm range and organic matter content for ten Illinois soils at four moisture levels -- oven dry and three others varying from 5% to 25%. Better correlations were obtained with information from the visible range (400 to 780 nm) than with near infrared (NIR) data (780 to 2400 nm). A first derivative model using optical density (OD, defined as $\log_{10}(1/\text{reflectance})$) data yielded an r^2 of 0.85 with the calibration dataset. Other researchers could not obtain satisfactory correlations when implementing this model with an experimental sensor and an expanded test set of 30 Illinois mineral soils [6].

Griffis [7] estimated the total carbon content of 18 air dry soils with an r^2 of 0.75 by using a phototransistor to measure the light reflected from soil illuminated by an incandescent lamp. Shonk and Gaultney [8] developed a real-time soil organic matter sensor intended to be calibrated for use within a given soil landscape, rather than for the larger geographic area (such as an entire state) attempted by other researchers. Laboratory tests yielded good correlations ($r^2 = 0.80$ to $r^2 = 0.98$) within a single landscape at a single moisture content. Field tests showed a curvilinear relationship between sensor output and organic matter, with independent calibrations needed for variations in travel speed or sensing depth.

Once obtained, the soil organic matter content signal could be used as a control input for a variable rate herbicide sprayer. Automatic control of application rates on the basis of travel speed is now commercially available, but the rate can be adjusted only over a narrow range. To effectively control application rate based on organic matter content, a system capable of achieving a maximum flow rate to minimum flow rate (turn down) ratio of 5:1 or greater is required [9].

This turn down ratio can be obtained by variation of either the herbicide concentration or the total nozzle output [10].

The variable concentration method allows spraying of a constant total volume of solution at a constant nozzle pressure, thus maintaining uniform coverage with currently available nozzle designs. Problems which have been noted with this method include increased deposition variability and transient errors in application rate resulting from time lags in the system [10].

Variations in total nozzle output can be accomplished by changing the inlet pressure to the nozzle, which may affect droplet size and distribution, or alternatively by the use of bypass nozzles which return a portion of their inlet flow back to the supply tank. Han et al. [9] developed and tested a laboratory model of a microprocessor controlled bypass nozzle sprayer system feasible for integration with a soil organic matter sensor. The application rate was controlled by regulating the pressure in the bypass line. Tests of the system showed it to have a quick, stable response to step input changes and to be capable of achieving a 5:1 turn down ratio.

To date, researchers have used a variety of data types and wavelength ranges to predict soil organic matter content using reflectance information. The data types have included color coordinates [1], broadband spectral reflectance [3,7], and spectral reflectance at one or more narrow-band wavelengths [5,6,11]. Wavelengths in both the visible [1,3,5,6,8] and the NIR [5,11] regions of the spectrum have been studied. No single combination of data type and spectral region has exhibited a consistently superior predictive capability.

This paper summarizes the results of a two-part project to develop and test an organic matter sensor which could provide control information for a variable rate herbicide sprayer [12,13,14]. First, the predictive capabilities of various reflectance and color based organic matter sensing methods were evaluated, and an optimum combination of data type, wavelength range, and calibration method was selected. Then, a prototype sensor was designed to implement the chosen method and tested both in the laboratory and in the field.

EVALUATION OF SENSING METHODS

Several sensing methods were evaluated, seeking a calibration which could be applied to a set of representative Illinois soils with a standard error of prediction (SEP, the standard error of the estimate in a validation dataset) of 0.29% carbon (0.50% organic matter) or less. Data in each of four standard color coordinate systems were analyzed using multiple regression techniques. Narrow-band spectral reflectance data in the visible and NIR ranges were also analyzed, using stepwise multiple linear regression (SMLR), principal components regression (PCR), and partial least squares regression (PLSR) [15].

PCR and PLSR are latent variable methods that reduce a set of collinear independent variables to a smaller set of orthogonal components which represent most of the variability in the original data.

In both methods, the optimum number of components to retain can be
determined by the technique of cross-validation. PCR consists of
first reducing the dimensionality of the data using principal component analysis (PCA) and then independently performing a multiple linear regression to relate the PCA factors to the dependent variable.
PLSR is similar to PCR, but the components extracted from the data are
a function of the values of both the independent and the dependent
variables. The reduction of data dimensionality and regression are
performed simultaneously so that small but relevant differences which
might be ignored in a PCR are included in the solution [15].

Test Procedures

Thirty Illinois mineral soils, selected to span the state's important agricultural soils, formed the calibration dataset. These
soil samples, which were also used in previous organic matter sensing
projects [5,6,16,17], were prepared for analysis by removal of foreign
material and crushing to pass through a 2 mm square mesh sieve.

Soil property determination: Visible and NIR reflectance curves
for the 30 soils were obtained with a research-grade spectrophotometer
at the USDA Instrumentation and Sensing Laboratory, Beltsville, MD
during a previous investigation [16]. Data were collected for three
replicate samples of each of the 30 soils at the 0.033 MPa (1/3 bar,
field capacity), 1.5 MPa (15 bar, wilting point), and air dry moisture
tension levels (n = 270). Two spectra were obtained for each sample:
visible/NIR data from 400.8 nm to 1116.8 nm on a 0.8 nm bandpass and
NIR data from 802.0 nm to 2592.0 nm on a 2.0 nm bandpass (Fig. 1).
The data were transformed to decimal reflectance (1.00 = a perfect
reflector) by comparison to a powdered halon standard. A Minolta
Model CR-110 Chroma Meter was used to collect color coordinate data
from a separate preparation of the 270 calibration soil samples.

Soil moisture content and texture were determined by standard
laboratory methods [16]. Total organic carbon was determined by dry
combustion in a LECO Model HF10[1] induction furnace. All calibrations
were done with organic carbon, rather than organic matter (1.72 times
organic carbon) as the dependent variable. The organic carbon contents of the 30 soils ranged from 0.45% to 3.16% (0.77% to 5.45% organic matter).

Data reduction and analysis: To bracket the soil moisture conditions likely to be encountered during field use of an organic matter
sensor, the 0.033 MPa and 1.5 MPa data were analyzed as a single
group. The air dry samples were not included in this analysis because
preliminary analyses indicated they would bias the regression toward
the drier end of the moisture range and reduce the predictive capability of the data.

---

[1] Trade names are used solely for the purpose of providing specific information. Mention of a trade name, proprietary product, or
specific equipment does not constitute a guarantee or warranty by the
U. S. Department of Agriculture and does not imply the approval of the
named product to the exclusion of other products that may be suitable.

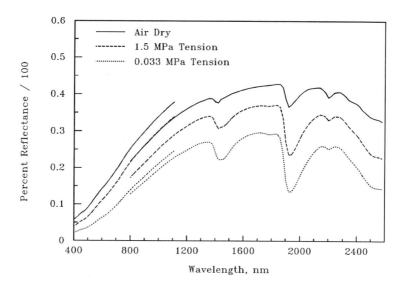

Figure 1. Mean spectral reflectance curves for Ade Loamy Sand at three moisture tension levels.

    Raw data obtained from the Minolta colorimeter in the form of chromaticity coordinates were transformed to three other color coordinate systems -- the tristimulus values and the CIELAB and CIELUV uniform color coordinates -- using standard methods [18]. Tristimulus values were also calculated from the spectral reflectance curves for the 270 soil samples by the weighted-ordinate method [18] and transformed to the other color coordinate systems. Multiple linear and quadratic regressions were performed to relate the color coordinates in each of the systems to organic carbon content for the combined 0.033 MPa and 1.5 MPa data.

    The visible/NIR spectral data were analyzed by SMLR, PLSR, and PCR. Previously developed SMLR prediction equations based on the 30 Illinois calibration soils [17] were applied to linearly independent validation datasets.

    Since soil color was thought to be a viable indicator of organic matter content, emphasis was placed on separate PLSR analysis of the visible data (400 to 780 nm on a 5 nm spacing). Analyses were done with the raw reflectance data and three transformed datasets:

    1. optical density (OD) transformation: $\log_{10}(1/\text{reflectance})$
    2. Kubelka-Munck transformation: $(1-\text{reflectance})^2/2*\text{reflectance}$
    3. square root transformation.

A supplementary PLSR analysis of the complete visible/NIR dataset (400 to 1110 nm on a 10 nm spacing) was also completed to see if extending the sensing range would provide a higher correlation. The NIR dataset (800 to 2580 nm on a 20 nm spacing) was analyzed with PLSR using raw reflectance data and OD transformed data. Analyses were also done on

various subsets of the NIR data to determine what reduction in predictive capability would be seen if fewer individual reflectances obtained on a wider bandwidth were sensed.

PCR was applied to the OD transformed visible and visible/NIR
data. In all PLSR and PCR analyses, cross-validation techniques [15]
were used to ensure the validity of the calibration equation. Outliers identified by the program were eliminated from the calibration
step to obtain the best possible fit; however, these outliers were
retained in the prediction step.

Results and Discussion

No calibrations based on data in the visible or visible/NIR range
achieved the desired SEP of 0.29% carbon or less. The best color
coordinate predictor of organic carbon was a quadratic regression
using CIELAB coordinates from the colorimeter, with an r^2 of 0.72 and a
standard error of the estimate (SEE) of 0.36% carbon. A quadratic
regression on CIELAB coordinates was also the best predictor when
using the spectral color coordinate data, with an r^2 of 0.62 and an SEE
of 0.42. The best valid SMLR equation, based on optical density
transformed visible data, had an r^2 of 0.81 and an SEP of 0.34% carbon.

Latent variable analyses of the optical density transformed visi-
ble/NIR data yielded slightly better results than did the SMLR analysis (PLSR: r^2 = 0.81, SEP = 0.32; PCR: r^2 = 0.76, SEP = 0.33). Analyses of the visible data alone were slightly less predictive. The
optical density transform was more predictive than the other transfor-
mations or the raw reflectance data, and partial least squares calibrations were more predictive than principal components calibrations.

The combination of the NIR (800 to 2580 nm) dataset and PLSR
yielded the best correlations between reflectance properties and organic carbon of any of the combinations tested. A maximum r^2 of 0.92
and a minimum SEP of 0.20% carbon were obtained with OD transformed
NIR data on either a 40 nm or a 60 nm spacing (Fig. 2). For both the
40 nm and the 60 nm datasets, the minimum range of wavelengths possible before the prediction degraded was similar. (Degradation in
predictive capability was defined as an increase in the standard error
of prediction of more than 0.01% carbon from that of the full-range
data.) For the data on a 40 nm spacing, the minimum wavelength range
without degradation was 1730 to 2370 nm (17 data points), with an SEP
of 0.21. For the 60 nm data, the range was 1720 to 2380 nm (12 data
points), with an SEP of 0.20.

The NIR spectra provided better predictions of organic carbon
than the visible spectra by partially compensating for the effects of
soil moisture variation. Spectral reflectance in the 1700 to 2000 nm
region is strongly correlated with soil moisture content [11] due to
the presence of a water absorption band (Fig. 1). Analysis of the PLS
output showed that the data points in this absorption band had the
most influence on the regression, thus allowing calibration for decreases in soil reflectance due to increased soil moisture.

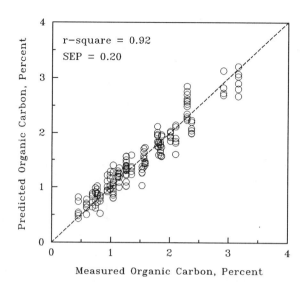

Figure 2.   Predicted vs. measured organic carbon obtained with PLSR from 0.033 MPa and 1.5 MPa moisture tension OD transformed spectrophotometer data in the NIR wavelength range.

SENSOR DESIGN AND TESTING

The overall design objective for the prototype soil organic matter sensor was to implement the chosen prediction method on a near real-time basis. Specific design objectives were:

1. A bandpass of 60 nm or less over a minimum sensing range from 1700 to 2420 nm, to implement the selected sensing method.
2. An essentially continuous (in wavelength) sensing method, to allow flexibility for additional optimization of the wavelengths selected for the prediction algorithm.
3. Potential, with additional refinements if needed, to acquire enough information to make a control decision every 4.5 seconds. (A 10 m (40 ft) spacing at a 2.2 m/s (5 mi/h) travel speed.)
4. Ability to predict soil organic matter content with an SEP of less than 0.5 percent (or less than 0.29% organic carbon) for the 30 Illinois soils in the calibration dataset.
5. Tolerance of dust, temperature fluctuations, shock loads, and vibration encountered in field operation.

The sensor which was designed to meet these objectives used a circular variable filter spinning at 5 Hz to sequentially provide monochromatic, chopped light with wavelengths of 1600 to 2900 nm from a broadband source. The monochromatic light was transmitted to the soil surface by a fiber optic bundle, allowing remote mounting of the major portion of the sensor. Energy diffusely reflected from the soil surface was captured by a lead sulfide (PbS) photodetector. The output from the detector was conditioned by an AC-coupled preamplifier

and input to the 80286-based personal computer used for data storage
and analysis through a Metrabyte DAS-16 data acquisition board.

Laboratory Test Procedures

    Initial tests of the sensor system documented its optical and
electronic performance characteristics and developed calibrations for
wavelength and reflectance level [12,14]. Once these initial tests
were completed, soil reflectance data were obtained.

    Soil sample preparation: The soil samples used to test the lab-
oratory operation of the sensor were subsamples of the same 30 Illinois mineral soils used to evaluate reflectance and color based sensing methods. Calibration soil samples, each consisting of approximately 150 g of dry soil, were prepared at the 1.5 MPa and 0.033 MPa
moisture tension levels.

    Properties measured for each soil included organic carbon, mois-
ture content, and cation exchange capacity (CEC) [12]. The organic
carbon measurements used in the prediction method selection process
were used again. The gravimetric moisture content of each sample was
determined by oven drying for 24 h at 105° C. The CEC was determined
for each of the soils using a procedure based on mechanical vacuum
extraction of the cations with an ammonium acetate solution at a pH of
7.0 and a subsequent Kjeldahl distillation [12].

    Organic carbon calibration: For the calibration tests, three
replications of the 30 Illinois soil samples at 0.033 MPa and 1.5 MPa
moisture tensions were analyzed using the laboratory sensor. Each
replication consisted of 60 randomized soil-moisture tension samples.
Approximately 40 g of soil were placed in a 50 mm diameter by 13 mm
deep flat black sample cup, and the surface struck off level with the
top of the cup. Each soil sample was placed under the sensor, and 100
multiple reflectance scans were obtained and stored. A ceramic disk
was scanned after every fifth soil sample and used as the reference
for the following five soil samples. The reflectance data were transformed to optical density (OD) and averaged to a 40 nm or 60 nm bandpass. Analysis of the averaged OD data was accomplished using PLSR.
The 0.033 MPa and 1.5 MPa moisture tension data were analyzed as a
single group to bracket the moisture conditions likely to be encountered in field operation.

    Prediction of other soil properties: The calibration set was
also used to predict soil moisture content and CEC. PLSR analyses
were used to correlate spectral information with laboratory determined
gravimetric moisture and CEC for the 30 Illinois soils.

    Height variation tests: A supplementary set of data was collect-
ed to study the effects of sample-to-sensor height variation on the
predictive capability of the organic carbon calibration. Samples of
six soils at 0.033 MPa moisture tension were used in the height variation tests. Three replications of data were collected at five distances between the sample and the quartz aperture window of the sensor: 5 mm, 10 mm, 15 mm (the nominal distance used for all other
tests), 20 mm, and 25 mm.

Moving sample tests: A sequence of tests assessed the prediction degradation associated with allowing a sample to move past the sensor head while collecting the reflectance scans. In this simulation of in-furrow field operation, the test soil was placed in a trough in a rotating disk within a light exclusion chamber. Two sample disk speeds were used, corresponding to linear velocities of 0.75 m/s and 1.5 m/s at the center of the sample trough. The six soils used in these tests were the same ones used in the varying height tests.

Laboratory Test Results

Organic carbon calibration: Comparison of the replicate spectral reflectance curves obtained with the prototype organic matter sensor and those obtained using the spectrophotometer at the USDA Instrumentation and Sensing Laboratory showed close agreement (Fig. 3). During the calibration tests, parameters were varied to find the optimum combination of wavelength range and scan averaging for organic carbon prediction. The best predictors of organic carbon were the 100-scan average datasets with either a 40 nm or 60 nm data spacing. The 40 nm data, consisting of 26 wavelength points from 1640 to 2640 nm, yielded an r^2 of 0.89 and an SEP of 0.23 (Table 1). The 60 nm data ($r^2 = 0.88$, SEP = 0.23) consisted of 17 wavelength points from 1670 to 2630 nm. Predictions made with 100-scan average data were significantly better than those made with 10-scan average data. In these analyses, as well as all other PLSR analyses, cross-validation techniques were used to ensure that the data were not overfit and to generate the standard error of prediction (SEP) values.

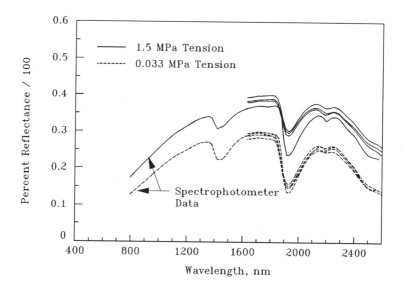

Figure 3. Comparison of mean spectrophotometer data and three replicate prototype sensor data curves for Ade Loamy Sand.

TABLE 1 -- Summary of PLSR prediction results obtained using the NIR organic matter sensor in the laboratory.

Dataset	R^2	SEC[a]	SEP[b]	CV[c]
Organic carbon				
0.033 MPa & 1.5 MPa				
10-scan average	0.85	0.26	0.28	18.9
100-scan average	0.89	0.22	0.23	15.4
Cation exchange capacity				
0.033 MPa & 1.5 MPa				
10-scan average	0.85	3.61	3.91	15.8
100-scan average	0.86	3.45	3.59	14.6
Soil moisture content				
0.033 MPa & 1.5 MPa				
10-scan average	0.96	1.58	1.69	11.8
100-scan average	0.97	1.48	1.59	11.1
0.033 MPa, 0.33 MPa, 1.5 MPa & air dry -- 100 scan	0.94	1.86	1.88	16.9

[a]SEC (standard error of calibration) is the standard error of the estimate in the calibration data, in units of pct. carbon, mEq/100g (cation exchange capacity), or pct. water.
[b]SEP (standard error of prediction) is the standard error of the estimate in the validation data, in the same units as SEC.
[c]CV (coefficient of variation) is the ratio of the SEP to the mean of the observed data, in pct.

Prediction of other soil properties: For CEC and soil moisture prediction, the 40 nm data (1640 to 2640 nm) were used, with 10-scan or 100-scan averaging. Limited comparisons with predictions using 60 nm data showed no appreciable difference in predictive capability from the 40 nm data. The best CEC prediction yielded an SEP of 3.59 mEq/100g for the combination of 0.033 MPa and 1.5 MPa moisture tensions. Moisture content was predicted with an SEP of 1.88 percent water for the dataset including 0.033 MPa, 0.33 MPa, 1.5 MPa, and air dry soil. In terms of the coefficient of variation (CV), the prediction of these two properties was more accurate than the prediction of soil organic carbon (Table 1). Thus, the sensor could be used to predict soil moisture and CEC as well as organic carbon.

Height variation tests: Visual observation of the reflectance data obtained in the height variation tests showed considerable offset between readings from the same sample at differing heights. In an attempt to reduce this variability, the decimal reflectance signal was normalized to a mean value of 1.00 for each sample curve. This normalization collapsed almost all of the height variability out of the reflectance curves, while retaining significant inter-soil variability, as needed for the prediction process. The PLSR analyses of the

varying height data showed that inclusion of calibration samples which
were scanned at various sample-to-sensor distances improved the capabilities of the sensor, even though normalization had removed the
majority of the observable difference between reflectance curves obtained at varying sample distances.

Moving sample tests: Data from the moving sample tests were
normalized prior to analysis to remove a portion of the sample-to-
sample variability seen in the reflectance curves. Inclusion of moving sample data in the calibration step yielded the best carbon predictions from this data, with an SEP of 0.38.

Sensor Modifications

The prototype soil organic matter sensor was modified for field
use (Fig. 4) and was operated to collect soil reflectance data under
field conditions. Modifications to the sensor system included the
addition of an opener to allow in-furrow sensing and modification and
packaging of the sensor to enable it to operate on a moving vehicle
and withstand field environmental conditions.

A horizontal disk row cleaner unit from a Hiniker Econ-O-Till was
chosen as the opener for the in-field tests of the soil sensor, since
it created a flat-bottomed furrow with uniform surface structure. The
sensor was contained in two sealed electrical enclosures attached to
the three-point hitch mounted row cleaner unit. Sensing of the furrow

Figure 4. NIR soil organic matter sensor configured for field
operation.

bottom was enabled through an aperture in the lower enclosure. Field
data acquisition was accomplished with the same DAS-16 analog and
digital I/O board used in the laboratory tests, installed in an expansion chassis attached to a Zenith Supersport laptop computer.

Field Test Procedures

Field tests were conducted at four locations on the University of
Illinois Agricultural Engineering Farm, Urbana, IL. At each location,
duplicate 90 m test runs were completed within 5 m of one another.
The tractor three-point hitch was allowed to float, such that the
depth of the sensor was controlled by the integral gage wheel. The
opener was adjusted to cut a 35 to 50 mm deep furrow.

Calibration of the sensor was a manual procedure, accomplished
with a test fixture which held a ceramic disk 15 mm from the sensor
head and excluded ambient light. A 50-scan reading of the ceramic
calibration standard was obtained before each test run, and a second
identical reading was obtained after the run was completed.

As each field test run was progressing, the location of the sensing head was marked every 10 s. Once the run was complete, a soil
sample of approximately 150 g was obtained at each marked location for
carbon and moisture analysis. Gravimetric moisture was determined by
oven drying for 24 h. Organic carbon was determined by dry combustion
in the LECO induction furnace. The mean of the 100 ceramic scans
associated with each run was used as the reflectance standard for the
run. In each field run the individual, unaveraged data scans were
converted to OD representation for PLSR analysis.

Field Test Results

Mechanical operation of the system in the field was quite good.
The furrow formed by the horizontal disk was generally smooth and
uniform. Isolated buried residue appearing in the path of the sensor
head and some roughness caused by adhesion of soil in the wettest
parts of the test runs were the only problems encountered. The average operating speed of approximately 0.65 m/s was selected since
bounce and pitch of the small tractor at higher speeds during preliminary tests had caused a rough, non-uniform furrow bottom.

Laboratory-determined carbon contents of the field samples ranged
from 1.36% to 3.42% (2.34% to 5.88% organic matter), as compared to
the 0.45% to 3.16% carbon range of the laboratory calibration samples.
The mean moisture content for the test samples was 19%. All samples
were in the range from 10% to 28% moisture, except for two samples
taken at the end of a run where the sensor did not penetrate past the
dry surface soil. Over 95% of the samples fell in the 1.5 MPa to
0.033 MPa moisture tension range of the calibration samples.

The calibration equation used to relate organic carbon to field
reflectance data was developed by applying PLSR analysis to normalized
100-scan mean laboratory data collected from 1640 to 2640 nm on a 40
nm bandwidth. The calibration dataset consisted of:

1. The standard calibration data; three replications of 30 Illinois mineral soils at each of the 0.033 MPa and 1.5 MPa moisture tension levels.
2. Varying height data; three replications of each of six selected soils at each of five sample-to-sensor distances, all at the 0.033 MPa moisture tension level.
3. Moving sample data; two replications of each of six selected soils at each of two sample velocities, all at the 0.033 moisture tension level.

These data were selected so that the main sources of reflectance variability (soil type, soil moisture, sample distance, and sample movement) were included in the calibration. The PLSR calibration with these data yielded a standard error of calibration (SEC) of 0.27% carbon and an r^2 of 0.82.

The calibration equation was applied to the individual reflectance scans within each field data run, generating a highly variable predicted carbon trace. The carbon data were processed through a 25-point (10 s) moving average routine after elimination of obvious outlier points with predicted carbon values of less than zero or greater than 6 percent. These averaged carbon traces were compared with the laboratory carbon measurements (Fig. 5) by extraction of predicted carbon values at the times corresponding to the laboratory samples. Good agreement was seen in the overall means (spectral mean = 2.44% carbon, laboratory mean = 2.41%, n = 102). The standard error of the means for the individual runs was 0.20% carbon. The overall SEP for the 102 observations was 0.53% carbon. This SEP was greater than the 0.50% standard deviation of the laboratory-determined carbon readings.

CONCLUSIONS

Of the methods evaluated, the combination of OD transformed NIR (800 to 2580 nm) spectral reflectance data and PLSR analysis was the most predictive of soil organic matter (r^2 = 0.92, SEP = 0.20% carbon). The other combinations tested failed to achieve the goal of a maximum SEP of 0.29% carbon (0.50% organic matter). The predictive capability of the NIR data was retained when the data were averaged to a 40 nm or 60 nm point spacing and the wavelength range was reduced to 1730 to 2370 nm.

The NIR organic matter sensor designed to implement the chosen prediction method was tested both in the laboratory and in the field. The sensor was able to predict organic carbon, CEC, and soil moisture in the laboratory, across a range of soil types and moisture contents. The organic carbon predictive capability of 100-scan averaged data obtained with the prototype sensor (r^2 = 0.89, SEP = 0.23% carbon) approached the predictive capability of data obtained on the same soils with a research-grade spectrophotometer.

During limited in-furrow field operation, the maximum SEP goal of 0.29% carbon was not achieved, due to movement of the soil past the sensor during the scanning process. Better predictions during field operation would likely require a means of holding the soil sample stationary during the scanning process. Although improved sample

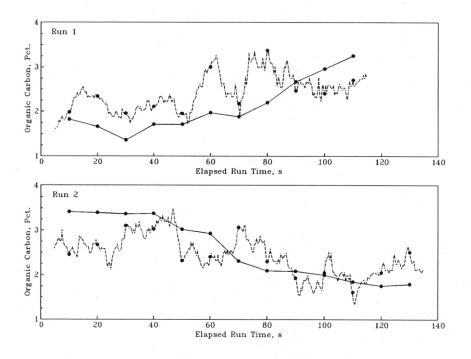

Figure 5. Measured (solid curve) and sensor predicted (dashed curve) organic carbon for field test runs 1 and 2. Data points indicate paired laboratory and sensor data used for statistical comparison.

presentation techniques would be required before the sensor could be used real-time in field operation, it could provide rapid, on-site organic matter determination for field mapping and subsequent chemical application rate control by means of a positioning system.

REFERENCES

[1] Page, N. R., "Estimation of Organic Matter in Atlantic Costal Plain Soils with a Color-difference Meter," _Agronomy Journal_, Vol. 66, 1974, pp. 652-653.
[2] Meggitt, W. F., "Herbicide Activity in Relation to Soil Type," in _Pesticides in the Soil: Ecology, Degradation, and Movement_, Proceedings of a Symposium, Michigan State University, East Lansing, 1970.
[3] Obukhov, A. I. and Orlov, D. S., "Spectral Reflectivity of the Major Soil Groups and Possibility of Using Diffuse Reflection in Soil Investigations," _Soviet Soil Science_, Vol. 2, No. 2, 1964, pp. 174-184.
[4] Baumgardner, M. F., Silva, L. F., Biehl, L. L., and Stoner, E. R., "Reflectance Properties of Soils," in _Advances in Agronomy, Volume 38_, Academic Press, Orlando, FL, 1985.

[5] Krishnan, P., Alexander, J. D., Butler, B. J., and Hummel, J. W.,
     "Reflectance Technique for Predicting Soil Organic Matter," Soil
     Science Society of America Journal, Vol. 44, 1980, pp 1282-1285.
[6] Pitts, M. J., Hummel, J. W., and Butler, B. J., "Sensors Utiliz-
     ing Light Reflection to Measure Soil Organic Matter," Transac-
     tions of the American Society of Agricultural Engineers, Vol. 29,
     No. 2, 1986, pp. 422-428.
[7] Griffis, C. L., "Electronic Sensing of Soil Organic Matter,"
     Transactions of the American Society of Agricultural Engineers,
     Vol. 28, No. 3, 1985, pp. 703-705.
[8] Shonk, J. L. and Gaultney, L. D., "Spectroscopic Sensing for the
     Determination of Organic Matter Content," ASAE Paper No. 88-2142,
     American Society of Agricultural Engineers, St. Joseph, MI, 1988.
[9] Han, Y. J., Bode, L. E., and Hummel, J. W., "Controlling Chemical
     Application Rate with Bypass Spray Nozzles," Transactions of the
     American Society of Agricultural Engineers, Vol. 29, No. 5, 1986,
     pp. 1221-1227.
[10] Hughes, K. L. and Frost, A. R., "A Review of Agricultural Spray
     Metering," Journal of Agricultural Engineering Research, Vol. 32,
     1985, pp. 197-207.
[11] Dalal, R. C. and Henry, R. J., "Simultaneous Determination of
     Moisture, Organic Carbon, and Total Nitrogen by Near Infrared
     Reflectance Spectrophotometry," Soil Science Society of America
     Journal, Vol. 50, 1986, pp. 120-123.
[12] Sudduth, K. A., "Near Infrared Reflectance Soil Organic Matter
     Sensor," Unpublished Ph.D. Thesis, Library, University of Illi-
     nois at Urbana-Champaign, 1989.
[13] Sudduth, K. A. and Hummel, J. W., "Optimal Signal Processing for
     Soil Organic Matter Determination," ASAE Paper No. 88-3038, Amer-
     ican Society of Agricultural Engineers, St. Joseph, MI, 1988.
[14] Sudduth, K. A., Hummel, J. W., and Funk, R. C., "NIR Soil Organic
     Matter Sensor," ASAE Paper No. 89-1035, American Society of Agri-
     cultural Engineers, St. Joseph, MI, 1989.
[15] Martens, H. and Naes, T., "Multivariate Calibration by Data Com-
     pression," in Near-Infrared Technology in the Agricultural and
     Food Industries, American Association of Cereal Chemists, St.
     Paul, MN.
[16] Worner, C. R., "Design and Construction of a Portable Spectropho-
     tometer for Real Time Analysis of Soil Reflectance Properties,"
     Unpublished M.S. Thesis, Library, University of Illinois at Ur-
     bana-Champaign, 1989.
[17] Smith, D. L., Worner, C. R., and Hummel, J. W., "Soil Spectral
     Reflectance Relationship to Organic Matter Content," ASAE Paper
     No. 87-1608, American Society of Agricultural Engineers, St Jo-
     seph, MI, 1987.
[18] Billmeyer, F. W. and Saltzman, M., Principles of Color Technol-
     ogy, 2nd Ed., John Wiley and Sons, New York, 1981.

William E. Hart and Larry D. Gaultney

DIRECT INJECTION OF DRY FLOWABLE AGRICULTURAL PESTICIDES

REFERENCE: Hart, W. E. and Gaultney, L. D., "Direct Injection of Dry Flowable Agricultural Pesticides", Pesticide Formulations and Application Systems: 10th Volume, ASTM STP 1078, L. E. Bode, J. L. Hazen, and D. G. Chasin, Eds., American Society for Testing and Materials, Philadelphia, 1990.

ABSTRACT: A prototype system was designed to inject dry flowable formulations directly into the hydraulic conduit of a liquid spraying system. Representative samples of dry flowable pesticides were obtained to evaluate parameters which affect liquid dispersion and metering characteristics of each material. Laboratory tests were conducted to quantify the dispersion times of the formulations at labeled rates. Preliminary tests indicated that the dispersion times for packaged formulations were lengthy, but reduced with increased amounts of agitation. Reduction of formulation particle sizes was investigated to determine effects on dispersion times. Tests were performed to evaluate the accuracy and repeatability of the metering unit with different pesticide formulations. A laboratory test apparatus was assembled to evaluate the overall performance of the system on an agricultural sprayer. Tests involved the collection of pump effluent to determine the compatibility of the unit with field sprayers equipped with hydraulic nozzles which can accommodate dispersible granular formulations.

KEYWORDS: direct injection, dry flowable pesticides, agricultural pesticide application, metering techniques, particle size reduction, pesticide dispersion times

The authors are: William E. Hart, Graduate Research Assistant, Agricultural Engineering Department, Purdue University, W. Lafayette, IN 47907; Larry D. Gaultney, Associate Professor, Agricultural Engineering Department, Purdue University, W. Lafayette, IN 47907.

INTRODUCTION

Background

Farmers in the United States used over 1.25 billion pounds of pesticides (herbicides, insecticides, fungicides, etc.) valued at over 6 billion dollars in 1985 [1]. Data from USDA indicates that corn was grown on 83.4 million acres and soybeans on 63.1 million acres. Almost 100% of these acres were treated one or more times with pesticides. Fruit and vegetable crops are also major farm products which require the use of significant amounts of pesticides.

Most pesticides are marketed as liquid formulations. Roughly 25 to 50% of these formulations are active ingredients, with the remainder consisting of water, emulsifiers, surfactants, and other adjuvants. Liquid formulations of pesticides have several limitations and risks. Liquids are expensive to transport due to the added weight of the carrier. They are susceptible to freezing and subsequent separation of emulsified components. Liquids are very difficult to clean up after inadvertent spills have occurred. Often, contaminated materials and soil must be handled as hazardous waste.

In addition to environmental risks, liquid pesticides also pose a significant risk to pesticide application personnel [2]. Splashback, drips, spills, and leaks often result in direct contact of concentrated pesticides on skin or clothing of farm workers [3]. Liquid formulations are often readily absorbed by the skin or may volatilize and be inhaled.

In order to respond to the need for safer pesticides, agricultural chemical companies have begun to develop improved formulations called dry flowables. These materials are porous, loosely formed granules which are composed of powdered active ingredients held together with binding agents. Dry flowables maintain their granular characteristics during transportation and handling and are easily pourable from containers. Dry flowables have many advantages over liquid formulations. They typically contain 50 to 90% active ingredients, thereby reducing bulkiness and the cost of transportation. Dry flowables are more easily collected and removed after spills than liquids. They leave no appreciable residues in containers; thereby allowing disposal by conventional means such as burying or placing in landfills. As with wettable powders, dry flowables are more chemically stable than liquids. They have the additional benefit of not being dusty; a major drawback of powders.

Problem

The handling of agricultural chemicals poses potential health risks to farmers and custom applicators. Transporting, pouring, and mixing liquid chemicals is especially risky due to splashing, dripping, and spillage onto skin or clothing. Leftover pesticide mixtures in sprayer bulk tanks must be disposed of, resulting in the introduction of chemicals into the environment. Rinsates from sprayer tanks are difficult or extremely expensive to dispose of properly. They are usually dumped on the ground at the rinse site or applied in a second pass over previously sprayed crops, resulting in unnecessary double coverage. Used chemical containers pose a significant risk for groundwater pollution when dumped in landfills, sink holes, or other typical disposal sites.

OBJECTIVES

Objectives of this study were to design and test a direct injection system that would accurately meter and mix dry flowable pesticides with liquid carriers in an agricultural sprayer.

PROCEDURES

Preliminary Laboratory Work

Representative samples of dry flowable pesticides were obtained to evaluate parameters which affect the liquid dispersion and metering characteristics of each material. Pesticides were chosen to include a representative range of active ingredient levels, particle sizes, and particle size distributions. Pesticides were limited to those currently used in typical agricultural practices.

Laboratory tests were conducted to measure the dispersion times of the formulations at labeled rates. Dispersion times were determined by mixing packaged formulations into water with a portable magnetic stirrer and visually observing the time required for the material to disperse. The magnetic stirrer was operated at its maximum speed (approximately 1000 rpm). Quantities of granulated pesticide were added to 100 ml of water to produce a concentration within the manufacturer's recommended application rates.

The particle size of selected pesticide formulations were physically altered to evaluate dispersion times as a function of the particle size. Pesticide formulations were crushed using a mortar and pestle. Dispersion times for the crushed material were evaluated using the same procedure as described for the packaged material.

Particle size distributions and bulk densities were determined for selected pesticide formulations. A portable sieve shaker conforming to A.S.T.M. E-11 specifications equipped with sieve numbers 10, 20, 40, 50, 70, 100, and 120 was used to segregate the granular material. The procedures outlined by Standard Test Method for Particle Size Distribution of Granular Carriers and Granular Pesticides ASTM E726, and Standard Method of Sampling Granular Carriers and Granular Pesticides ASTM E725 were followed to sieve 100 gram samples for one minute tests. Bulk densities were determined using a modification of Standard Test Methods for Determining Bulk Density of Granular Carriers and Granular Pesticides ASTM E727, in which a smaller sampling container was used with the omission of a filling hopper. The smaller container was required to allow for multiple test replications on limited quantities of selected pesticide compounds.

Apparatus Description

A prototype system was designed and assembled to continuously meter and mix dry flowable pesticide formulations with a liquid carrier. Primary components of the system included a metering/crushing unit and an interface to inject dry pesticide materials into the liquid carrier.

The metering/crushing unit was designed to to perform two functions: 1) meter granular material and 2) reduce particle size. The unit consisted of a screw housed inside a constant bore sleeve. Screw design incorporated a tapered shaft with constant diameter lands. All components of the unit were constructed from stainless steel with the exception of some bearing surfaces. Figure 1 shows a two dimensional illustration of the metering/crushing screw.

The solid-to-liquid interface was constructed from plexiglas to allow visual observation of the mixing chamber. The unit is illustrated in Figure 2. A two-piece design allowed the interface to be clamped around the screw housing to provide a water-tight seal. Inlet and outlet ports were equipped with female pipe threads to accept hose barb connectors, thus allowing inline installation along the suction supply line.

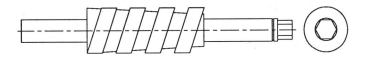

Figure 1. Tapered screw designed to meter and crush granular pesticides.

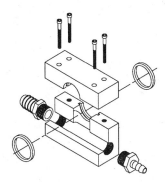

Figure 2. Interface designed to mix metered pesticides with liquid carrier.

The compact design of the metering/crushing unit and the solid-to-liquid interface allow the unit to be installed on conventional agricultural sprayers with few modifications. This feature permits sprayer systems to be used as a direct injection system, as a conventional tank mix systems or as a combination of both systems. An exploded view of the direct injection system is illustrated in Figure 3.

Metering Test Setup

The metering/crushing screw was rotated by a vane-type hydraulic motor connected to a stationary power supply. Infinite speed control was accomplished with a pressure compensating flow control valve on the motor supply line. An inline hydraulic flow tester was used to monitor system parameters such as flow, pressure, and temperature. Screw shaft rotation was measured using a proximity switch. Four metallic transition points were installed on the shaft couple to provide inputs to the proximity switch. The screw housing sleeve was equipped with a copper-constantan (Type-T) thermocouple to measure sleeve surface temperature. A Campbell Scientific Inc. 21X Micrologger was used to read and record all

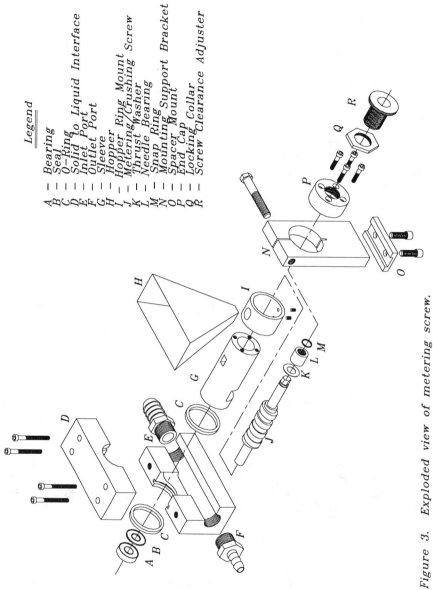

Legend

A — Bearing
B — Seal
C — O-Ring
D — Solid to Liquid Interface
E — Inlet Port
F — Outlet Port
G — Sleeve
H — Hopper
I — Hopper Ring Mount
J — Metering/Crushing Screw
K — Thrust Washer
L — Needle Bearing
M — Snap Ring
N — Mounting Support Bracket
O — Spacer Mount
P — End Cap
Q — Locking Collar
R — Screw Clearance Adjuster

Figure 3.   Exploded view of metering screw.

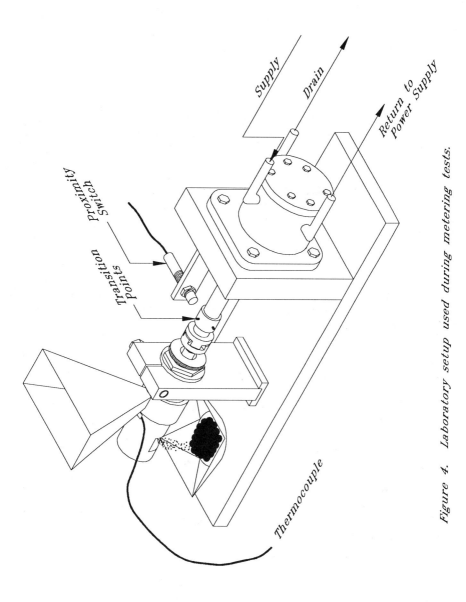

Figure 4.   Laboratory setup used during metering tests.

sensor outputs. During metering tests, the solid-to-liquid interface was not connected to the system. Refer to Figure 4 for a pictorial representation of the system used.

Data Collection

Metering tests consisted of collecting timed output samples from the metering/crushing unit discharge. Samples were weighed with a digital balance to the nearest 0.01 gram. All sensor outputs were read and recorded by the Campbell Micrologger during collection of discharge samples. At initial startup the system was run empty to allow the hydraulic power supply to reach its normal operating temperature; thereby reducing fluctuations in speed due to changes in oil temperature and viscosity. Tests were performed over a speed range of 100 to 350 revolutions per minute (rpm). Screw speed settings were confirmed with a hand held tachometer at no-load conditions. Tests were replicated a minimum of three times at each speed setting. The screw was run under load for approximately two minutes at each speed setting prior to data collection to allow the system to reach a steady-state condition. Screw output was evaluated at one and three minute collection periods.

Injection Test Setup

The solid-to-liquid interface was attached to the metering/crushing screw to evaluate the performance of the unit in mixing pesticide concentrations with a liquid carrier. A laboratory scaled sprayer system was assembled to meter concentrated pesticide at a point upstream of the system pump as illustrated in Figure 5. A Hypro Model D19 twin-diaphragm pump was used for the system pump. The diaphragm pump was driven with an electric motor through a belt drive. By-pass fluid from the pressure regulating valve was plumbed to the low pressure, suction inlet to the system pump as described by Tompkins and Howard [4]. This recirculating loop allowed for some additional mixing of pesticide concentrates with the liquid carrier. System operating pressure was monitored with a pulse dampened gauge downstream from the pressure regulator. The scale of the metering/crushing unit provided sufficient output to supply a two nozzle boom. Two Spraying Systems 8004VS Teejet flat fan nozzles with 50-mesh strainers were installed on the boom section. Also, a line strainer (equipped with a 50-mesh screen and clear bowl) was plumbed inline between the system pump and the boom section.

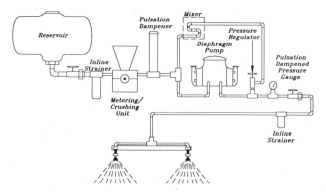

Figure 5.   Sprayer system used in laboratory injection tests.

RESULTS AND DISCUSSION

Dispersion Times

Packaged formulations: Tests indicated that the times required to disperse packaged formulations into water were lengthy, but reduced with increased amounts of agitation. Average dispersion times for selected pesticides are presented in Table 1. Average times were calculated from four replicated samples. Dispersion times ranged from 24 to over 50 seconds. Examination of these results strongly suggested that direct injection of packaged materials would be impractical and in most cases infeasible.

TABLE 1 -- Dispersion times for 187 liter per hectare (20 gpa) equivalent mixture.

Product Name	Application Rate kg/ha (oz/ac)	Equivalent Rate grams/100 ml	Time (seconds)
Lorox Plus	0.98 (14)	0.52	41
Gemini	0.98 (14)	0.52	44
Lexone DF	0.56 (8)	0.30	51
Preview	0.56 (8)	0.30	24

Physically altered formulations: Pesticide formulations crushed with the mortar and pestle were evaluated in a manner similar to that of packaged formulations. Crushed pesticide formulations dispersed almost immediately upon contact with water. Exact dispersion times were not recorded, but it was evident that material of this form could be introduced directly into a liquid conduit.

Particle size distributions and bulk density measurements for five products are presented in Table 2. Reported results are mean values for replicated test samples. It should be noted that over 85% of the particle diameters were within a range of 425 to 2,000 μm for all of the pesticides evaluated. Bulk densities were consistent within each formulation, but varied substantially between pesticide formulations. For example, a 43% increase in bulk density was noted between AAtrex Nine-O and Lexone DF.

Metering/Crushing Unit Design

A prototype direct injection system was designed based on the results from preliminary laboratory tests. Design considerations included the following:

- a laboratory scaled unit that would allow the evaluation of direct injection without requiring large amounts of pesticide formulations,
- a mechanism that would provide particle reduction through a combination of shear and compression forces,
- a feed mechanism that would accommodate the range of particle sizes without bridging problems, and finally

- a unit that was highly adjustable for speed and particle size reduction since the effect of several factors were still unknown.

TABLE 2 -- Particle distributions and bulk densities for selected pesticide formulations.

Diameter Ranges	Product Name				
	Lorox DF	Lexone DF	AAtrex	Bladex	Harmony
2.00 mm < D	0.00	0.00	9.04	3.67	0.00
850 μm < D < 2.00 mm	28.65	46.14	78.68	86.46	23.92
425 μm < D < 850 μm	63.35	40.19	11.34	9.68	62.53
300 μm < D < 425 μm	7.81	8.89	0.53	0.17	13.12
210 μm < D < 300 μm	0.19	3.49	0.21	0.02	0.39
150 μm < D < 210 μm	0.00	1.08	0.14	0.00	0.03
D < 150 μm	0.00	0.21	0.06	0.00	0.01
Bulk Density g/cm^3 (lbs/ft^3)	0.54 (33.71)	0.66 (41.20)	0.46 (28.72)	0.52 (32.46)	0.63 (39.33)

Metering and Size Reduction Evaluation

Initially, the metering/crushing unit was evaluated for its compatibility with several pesticide formulations. Small samples of selected pesticides were metered through the unit for short time durations to determine whether the material would or would not flow through the unit. Lorox Plus, Lorox DF, AAtrex Nine-O, and Manzate 200 DF flowed through the unit extremely well. Bladex 90 DF and Harmony worked well, but showed some indications of resistance. Preview, Lexone DF, and Gemini flowed poorly and in some instances "locked" the unit. Since the primary focus of the study was to evaluate the feasibility of using direct injection technology with dry flowable pesticides, it was decided to study the initial performance using the pesticide formulation Lorox DF.

The metering/crushing unit was evaluated to determine its ability to meter predictably and reduce formulation particle sizes. Tests included operating the unit over a speed range of 100 to 350 rpm (no-load) at three screw position settings. Screw position settings refer to the relative position of the metering/crushing screw in relation to the screw sleeve, thus determining the discharge clearance. Settings will be referred to as A, B, and C, with clearances of 0.46 mm, 0.42 mm, and 0.38 mm, respectively.

The screw was found to meter reliably between the range of 3.9 and 19.8 grams per minute using Lorox DF. Replicated tests were repeatable and predictable. One and three minute data collection periods were evaluated. Statistical analysis determined that data collected at the two time intervals were not significantly different.

Screw performance for particle reduction and metering was determined to be a function of speed and screw position for a particular formulation. The ability of the screw to reduce packaged material particle size is illustrated in Table 3. Setting A was able to reduce the percentage of material whose diameters were larger than 425 μm from over 90% to approximately 50% for Lorox DF. Settings B and C further reduced this percentage to approximately 35 and 27%, respectively. The small percentages remaining in the 850 - 2,000 μm range for all three settings were extremely fine particles which collected on the screen and would not respond to the shaking action.

TABLE 3 -- Particle distributions for Lorox DF at three screw settings.

Diameter Ranges	Particle Distributions			
	Initial	Setting A	Setting B	Setting C
2.00 mm < D	0.00	NA[b]	NA[b]	NA[b]
850 μm < D < 2.00 mm	28.65	0.29	0.32	0.23
425 μm < D < 850 μm	63.35	49.86	34.99	26.64
300 μm < D < 425 μm	7.81	20.01	20.90	25.80
210 μm < D < 300 μm	0.19	5.89	8.34	9.84
150 μm < D < 210 μm	0.00	4.58	5.59	8.99
125 μm < D < 150 μm	0.00	4.29	4.96	6.75
D < 125 μm	NA[a]	15.08	24.90	21.75

[a]Note: Sieve number 120 not used on 6-sieve shaker.
[b]Note: Sieve number 10 not used on 6-sieve shaker.

Metering characteristics for the unit using Lorox DF are presented in Figures 6 and 7. Screw output increased as the speed of operation increased (Refer to Figure 6). There was a slight difference in output due to screw setting at comparable speeds. These data were also analyzed on the basis of output per revolution (Figure 7). Both figures 6 and 7 show that output at the three settings are almost equal and follow basically the same trends. Output exhibits slight fluctuations at lower operating speeds but fluctuations tend to dissipate as speed increases.

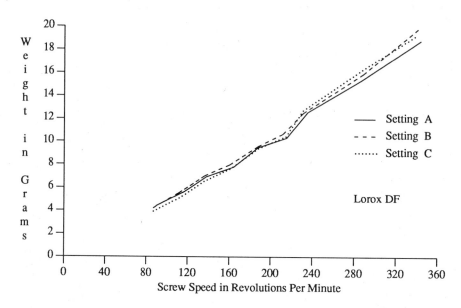

Figure 6. Metering rates as a function of speed for three screw settings.

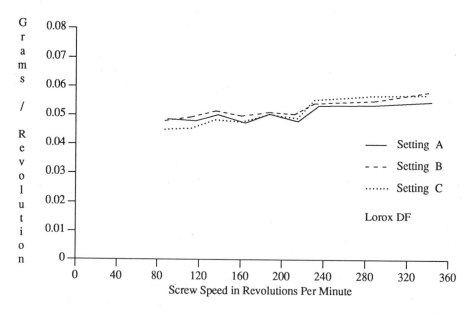

Figure 7. Metering flow rates as a function of rpm for three screw settings.

Sleeve temperatures were recorded during metering tests. A stable operating temperature was noted for each speed and pesticide combination. Visual inspection of metered material indicated output textures ranging from a fine dust-like powder to a coarse ribbony composition. It was further noted that this textural change was formulation dependent. Since it was not apparent that the change was strictly due to speed and/or screw position settings, a test was initiated to determine the effects of temperature on screw output. Results of those tests are presented in Figures 8 and 9. Sleeve temperature was dependent on operating speed and screw setting. For a given set of operating parameters, a steady-state condition was always obtained. Metering unit discharge rate was affected by temperature initially, however a near-constant temperature was reached within a few minutes of operation.

Formulation delivery to the metering/crushing unit was extremely successful with the prototype hopper and inlet design. No bridging problems were experienced with the hopper unit. There were no measurable differences in output due to depth of material in the hopper. Slight modifications to the inlet and hopper base may be required to accommodate other materials with larger particle sizes.

The metering/crushing unit was used successfully to meter Lorox DF into the sprayer system illustrated in Figure 5. A rate of 9.4 grams per minute was mixed with a water flow rate of 3.03 l/min, which equates to an application rate of 0.69 kg/ha (0.62 lbs/ac). No traces of pesticide residue were found on the 50-mesh strainer screen downstream from the pump or on the 50-mesh nozzle strainers. The 8004 flat fan nozzles operated without any difficulties for the duration of all tests.

SUMMARY AND CONCLUSIONS

A study was initiated to investigate the feasibility of utilizing direct injection technology with dry flowable pesticides. Preliminary tests indicated reduction of packaged pesticide formulations particle sizes decreased dispersion times. A meter/crushing unit was designed and constructed to reduce formulation particle size, meter the formulation, and introduce the formulation to a hydraulic conduit. Laboratory tests indicated that metering and crushing were consistent, repeatable, and successful in reducing pesticide formulation to a quickly dispersible particle size. The unit was used to introduce the pesticide Lorox DF at a rate of 0.69 kg/ha into a laboratory scale agricultural sprayer. The sprayer operated reliably using 8004 flat fan nozzles with 50-mesh screens.

ACKNOWLEDGEMENTS

The authors gratefully acknowledge the help of Jesse Davis for his technical assistance with graphics, The University of Tennessee Agricultural Engineering Department (Knoxville, Tn) for parts fabrication, and John Nickle, E. I. Dupont and Gene Chamberlain, Dow-Elanco for supplying chemical formulations and technical asssitance.

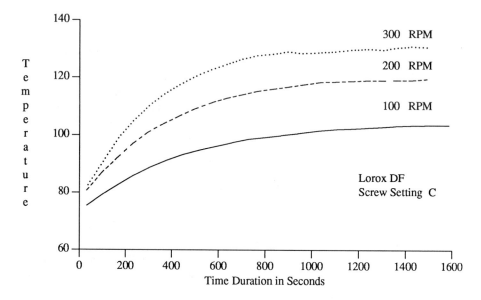

Figure 8. Metering sleeve temperature as a function of time at three operating speeds.

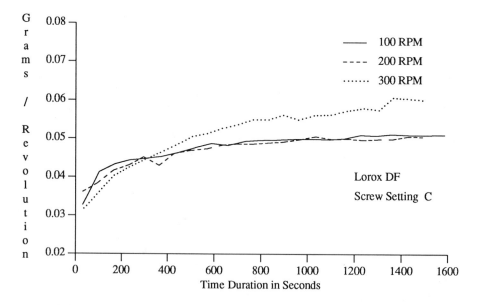

Figure 9. Metering unit discharge rate reaching steady-state flow at three operating speeds.

REFERENCES

[1] Agricultural Statistics, U.S. Department of Agriculture, Washington, D. C., 1987.

[2] Putnam, A. R., Willis, M. D., Binning, L. K., and Boldt, P. F., "Exposure of Pesticide Applicators to Nitrofen: Influence of Formulation, Handling Systems, and Protective Garments", Pesticide Research Center, MSU, East Lansing, MI, 1983.

[3] Muir, N. I. and Gover, G., "Pattern of Dermal Deposition Resulting from Mixing/Loading and Ground Application of 2.4-Dimethylamine Salt", Second Int'l Conference on Performance of Protective Clothing, ASTM, 1987.

[4] Tompkins, F. D. and Howard, K. D., "Boom Characteristics with Direct Chemical Injection," Paper No. 88-1593, American Society of Agricultural Engineers, St. Joseph, MI, 1988.

Alton O. Leedahl and Glen L. Strand

ON-THE-G0 APPLICATION OF DRY AND LIQUID HERBICIDES WITH GRANULAR FERTILIZER

REFERENCE: Leedahl, Alton O. and Strand, Glen L., "On-the-Go Application of Dry and Liquid Herbicides with Granular Fertilizer," PESTICIDE FORMULATIONS AND APPLICATION SYSTEMS: 10th Volume, ASTM STP 1078, L. E. Bode. J. L. Hazen, and D. G. Chasin, Eds., American Society for Testing and Materials, Philadelphia, 1990.

ABSTRACT: On-the-go application of dry and liquid herbicides with dry granular fertilizer is a successful commercial reality. Liquid herbicide is coated on dry granular fertilizer in a patented process in the boom tube of a pneumatic applicator. The process takes a fraction of a second, it greatly reduces herbicide contamination of handling equipment and simplifies clean up. Dry granular herbicides are metered individually into each boom tube and blended with dry granular fertilizer on-the-go.

KEY WORDS: On-the-go, application, blending, metering, dry herbicides, liquid herbicides, coating, impregnation, granular fertilizer, variable rate, reduced contamination, clean-up.

The application of pesticides (pre-plant herbicides) with dry granular fertilizer is done in 3 different ways today.
1. Impregnation (or coating) of liquid pesticides on dry granular fertilizer in a blending operation.
2. On-the-go impregnation (or coating) of liquid pesticides on dry granular fertilizer.
3. On-the-go application of dry pesticides blended with dry granular fertilizer.

In the first method, concentrated liquid pesticides are coated (or impregnated) on dry granular fertilizer in the blender. Then the blended product is conveyed into a tender, transported out to the field and conveyed into an applicator. Then the blended product is applied on the field. In this process the tank of the applicator is contaminated with the pesticide coated fertilizer as well as the metering mechanism and boom. The tender is also contaminated with pesticide coated fertilizer as well as the conveyor and blender. Cleanup of all these pieces of equipment is a problem as well as disposal of residue from cleanup. Therefore a method of on-the-go coating of pesticides on dry granular fertilizer was developed.

Alton O. Leedahl is Manager of Engineering, and Glen Strand is Service and Product Manager at Tyler Limited Partnership, Box 249, Benson, MN, 56215.

In this new system [1] a novel process occurs in the boom of the unit. The natural principles of fluid dynamics are utilized to coat liquid pesticides on to dry granular fertilizer in the boom tubes. In this system the concentrated liquid pesticide is metered into each boom tube. The liquid clings to the inside wall of the tube and tends to coat the inside wall of the tube. The dry granular fertilizer is forced, by the high velocity air, to travel along the inside wall of the boom tube also. As the fertilizer moves it wipes the liquid off the tube wall and in this process the concentrated liquid pesticide is coated (or impregnated) on the dry granular fertilizer. On the shorter tubes in the boom over half of the liquid is coated on the fertilizer within the first 1/2 meter. Nearly all of the liquid is coated on the fertilizer within 1 meter.

In this system a separate peristaltic pump is used for each boom tube (20 pumps total). The pumps can be shut off in increments to give swath widths of 100%, 75%, 50%, or 25% of total boom width. The concentrated liquid pesticide is carried in two 190 liter (50 gallon) tanks located at the rear of the applicator. The liquid is drawn out of one of the tanks and pumped by the peristaltic pumps to outlets on each boom tube.

It is very important to note that the liquid pesticide does not come into contact with fertilizer metering mechanism, hopper or auger. Therefore the fertilizer hopper, auger and metering mechanism do not become contaminated with the pesticide. Near the end of the field the operator switches over to flush and less than 2 liters of water from the fresh water flush tank is used to replace the liquid pesticide in the pumps and lines. Then the pumps are run another 15 to 30 seconds using approximately 4 liters of water to flush out the lines.

A pressure washer option is available allowing a person to clean dust, etc. off the entire vehicle right out in the field. This eliminates the need for expensive wash pads, rinsate tanks and disposal of rinsate.

Now let's talk about the application of dry granular pesticides with dry granular fertilizer. In Tyler's system fertilizer is metered by using conveyor belts and dividers over the belts. The belts move from the center of the unit to each side. The fertilizer is metered individually into each boom tube as the fertilizer passes through a gate and then a series of dividers. Dry granular pesticide boxes are then mounted directly above or along the side of the dividers. A separate metering wheel is used between each divider to meter pesticides into each individual boom tube. In this way both fertilizer and pesticide are individually metered and then blended into each individual boom tube. The boxes have a total capacity in excess of 565 liters and can be used over a wide range of application rates.

In the spring/summer of 1989 a prototype of a commercial version of the organic matter probe developed by Purdue University [2,3] was tested on a Tyler pneumatic applicator. The probe is used for on-the-go application rate changes of herbicides according to soil organic matter content. Tyler has acquired the license from Purdue to use this technology and is in the process of developing a production unit which will be in limited production in 1990.

REFERENCES:
[1] Leedahl, Alton O. and Strand, Glen L., "Application of Pesticides On-the-Go with Granular Fertilizer," PESTICIDE FORMULATION AND APPLICATION SYSTEMS: INTERNATIONAL ASPECTS: 9th Vol., ASTM STP 1036, J. L. Hazen and D. A. Hovde, Eds., American Society for Testing and Materials, Philadelphia, 1989.
[2] Gaultney, L. D., et al. "Automatic Soil Organic Matter Mapping," Paper No. 88-1607, American Society of Agricultural Engineers, St. Joseph, Michigan, 1988.
[3] Shonk, J. L. and Gaultney, L. D., "Spectroscopic Sensing For The Determination of Organic Matter Content," Paper No. 88-2142, American Society of Agricultural Engineers, St. Joseph, Michigan, 1988.

Brian W Young

DROPLET DYNAMICS IN HYDRAULIC NOZZLE SPRAY CLOUDS

REFERENCE : Young, B. W., "Droplet Dynamics in Hydraulic Nozzle Spray Clouds", Pesticide Formulations and Application Systems : 10th Volume ASTM STP 1078, L. E. Bode, J. L. Hazen, and D. G. Chasin, Eds., American Society for Testing and Materials, Philadelphia, 1990.

ABSTRACT : The spraying of a pesticide through a traditional hydraulic nozzle produces a heterogeneously sized droplet cloud. Three of the factors determining the fate of the droplets subsequent to their formation are : (1) the mechanics of the atomization process itself, (2) the forward motion of the nozzle and droplet cloud, and (3) any ambient wind. The Particle Measuring Systems Inc. (PMS) Imaging Spectrometer has been used to characterize the droplet cloud; and this data is related to the behaviour observed during motion, when twin inwardly curling vortices are formed behind the nozzle. These vortices become susceptible to drifting away from the target. An attempt has been made to characterize the vulnerability of a particular nozzle to produce a driftable component by assessing a factor termed the Drift Potential. Data are presented for a range of nozzles.

KEYWORDS : hydraulic nozzle, droplet dynamics, PMS Spectrometer, drift potential, nozzle movement, vortices.

INTRODUCTION

Despite the many technical advances in sprayer design, construction and control, the heart of the spraying process remains the nozzle itself. For the majority of ground based vehicular applications the flat-fan hydraulic nozzle is still used and its design is essentially unchanged over decades. New concepts have emerged, but as yet are a minor part of the market. Much has

Technical Officer, Spray Application Unit, ICI Agrochemicals, Jealott's Hill Research Station, BRACKNELL, Berkshire, England, RG12 6EY

been written over the years about optimal application techniques and concepts, and much has been learnt also, however in most cases the farmer or contractor has to achieve a sub-optimal level of success because of practical limitations.

Many man-years of effort have researched, and many papers have set out to describe, the technical performance of the intrinsically simple device - the nozzle itself. Most fan pattern nozzle tips are a formed block of metal, plastic or ceramic with a single axial hole and an elliptical, external aperture. Sophisticated laser-based optical systems are now used (and indeed used for the work described in this paper) to characterize the droplet spectra produced from such nozzles [1] and computer modelling concepts are now being used to theoretically simulate the behaviour[2].

The basis of this paper is however to attempt to utilise some of these modern technologies to explain what can be observed to happen in a normal spraying operation. It has come from extensive experimentation and observation and a desire to understand "what actually happens" when a spray nozzle is passed over a target. Originally this was related to the behaviour in indoor track sprayers, but has now been extended to the outdoor situation. Indeed the concepts that will be outlined in this paper, take a particular relevance at present with the ever increasing expression of environmental concern about application technology and the reduction of spray drift.

The paper is divided into three sections: (1) The characterisation of the spray cloud, (2) The characterisation of the trailing plume, (3) The quantification of the trailing plume.

BACKGROUND

One standard procedure at Jealott's Hill for the evaluation of candidate chemicals is to spray potted plants or seed trays in a spray chamber using a single overhead, moving nozzle - usually a Spraying Systems 8001E flat-fan nozzle. Such nozzles are designed to give a more uniform swath over the fan width than the normal design. However, experience has shown that on many occasions the cross-sectional pattern of the deposited swath - as judged by using a tracer dye onto white card for example - shows a twin banding at the edges with a central region of lower intensity. The magnitude of this effect is a function of the operational parameters for a particular nozzle.

Why is this - and why do nozzle manufacturers characterize nozzle performance on a static patternator if the profile changes in motion? One question of course for such situations is should the nozzle move over the target, or should the target be moved under a stationary nozzle? This question is being addressed at present. This work has shown that the cross-sectional deposit profile (on a flat surface) under a single moving nozzle (such as an 8001 nozzle tip at 300kPa pressure at 1m/sec travel speed) is more uniform from a conventional flat-fan nozzle than an "Even" nozzle.

While it is well known that a moving nozzle leaves a trailing wake or plume, work in a research track sprayer using back-lighting techniques has shown that the trailing plume is in fact formed from inwardly curling vortices formed on either side of the fan as it moves along. This effect is most apparent when viewed from behind. As soon as a stable spray cloud from a stationery nozzle is moved, the cloud begins to form a trailing cusp shape. It is clear that the individual droplet trajectories become disturbed by the lateral motion. The degree of trajectory change is a function of individual droplet energies. Since the droplet cloud is a heterogeneously sized population the resultant plume is the combination of the individual effects. It is also influenced by the effect of the displaced airflow passing round thespray cloud. Interestingly independent studies at AFRC Silsoe (using smoke plumes in a wind tunnel) have indicated that very little air passes though a typical fan spray cloud but flows over the edges and curls round behind [3]. Studies by the author on a wide range of hydraulic nozzles have led to the conclusion that the formation of vortices is an inevitable consequence of movement of the spray cloud. The magnitude of the effect does of course depend on the individual operating parameters. Hollow cone, full cone and anvil type nozzles also demonstrate the effect; however the present work has been restricted to flat-fan nozzles for convenience.

A paper by Professor Gohlich at the 1979 CIGR Symposium in East Lansing [4] and a subsequent publication [5] have been valuable in supporting this concept. Although there is no direct reference to vortex formation they are apparent in one figure. An article in Agriculture International [6] shows the formation of such vortices from a conventional nozzle, and reference is made to "forward speed vortexing" in the publicity literature for the Airtec nozzle [7]. Of particular significance in verifying this effect has been a collaborative project with Hardi International in Denmark; where extensive filming (under back-light conditions) has confirmed the present observations on moving spray clouds. Given that the process of vortex formation is an aerodynamic effect it is interesting how closely (at least visually) the behaviour resembles the vortex formation during aerial spraying.

With the availability of modern instrumentation for characterizing sprays it is perhaps also surprising that little data has been published on the spatial variability of droplet size and velocity within a spray cloud -generally measurements are made in a single position. Limited data illustrating the significance of distance from the nozzle on the droplet velocity (and hence kinetic energy) has been published by the author [8]. The classic publication by Roth and Porterfield some 25 years ago gives very similar data [9]. Extensive high speed cine studies of spray clouds were made by Bode and colleagues [10].

The initial experimental work described here was therefore intended to 'map-out' the droplet size and velocity profiles throughout the spray cloud from a single stationary flat-fan nozzle. These profiles would then be used to aid interpretation of the effects observed when such a cloud was laterally displaced.

CHARACTERISATION OF THE SPRAY CLOUD

Experimental Method

Nozzle mounting: The single nozzle was mounted on an adjustable tilt-assembly, such that the nozzle could be oriented to ± 50° either side of vertically downwards. The longitudinal axis of the fan was tilted, enabling the portion of spray being measured to be travelling vertically downwards to the laser beam. The water was supplied to the nozzle by a typical diaphragm pump; a pressure gauge immediately before the nozzle monitored the applied pressure.

Laser mounting: The Particle Measuring Systems Inc. 2D Optical array spectrometer OAP-2D-GA1 was mounted vertically directly below the centreline of the nozzle. The probe resolution was set to cover the size range 15-900 micrometres. It was orientated such that the laser beam passed through the short axis of the spray cloud -ie. transverse to the fan. The data was collected using the PDPS-11C data system (with version 14 software); and transferred to a custom software package on an IBM-XT.

Sampling matrix: The nozzle was positioned above the laser probe such that a series of measurements were taken down the centreline of the spray cloud at distances of 20, 30, 40 and 50cm below the nozzle. At each point data were obtained at 200kPa and 400kPa pressure. The nozzle was then tilted such that the above sequence was repeated at each of 10°, 20° 30° and 40° off centre. The fan was considered to be symmetrical and only one side was measured. In all cases the measured spray vertically intersected the laser beam.

Results

The data presented here are for a Spraying Systems Inc. 8002 flat-fan nozzle tip, as representative of typical agricultural usage. The matrix described above clearly provides far more data than can be totally included here, so an attempt has been made to present the significant features.

TABLE 1 -- The effect of spatial position on the droplet size
distribution for an 8002 nozzle at 200kPa pressure

| Position | | Median Diameter | | Volume % less than | |
Angle	Distance (cm)	Number	Volume	110 microns	210 microns
0°	20	66	269	7.4	30.0
	30	74	262	8.1	32.2
	40	83	250	7.9	35.9
	50	86	246	6.4	36.1
10°	20	85	276	6.3	29.2
	30	95	275	5.7	28.2
	40	97	258	5.0	31.2
	50	96	255	4.9	32.6
20°	20	99	280	4.9	27.6
	30	122	273	3.2	26.5
	40	133	269	2.4	26.0
	50	141	265	2.2	25.3
30°	20	140	288	2.7	23.5
	30	186	299	0.7	17.1
	40	204	291	0.5	15.9
	50	214	289	0.2	15.1
40°	20	197	307	0.5	14.8
	30	218	314	0.2	11.9
	40	227	306	0.2	10.7
	50	248	314	0.0	7.0

TABLE 2 -- The effect of applied pressure on the droplet size
distribution, at a distance of 40 cm from the nozzle

| | | Median Diameter | | Volume % less than | |
Angle	Pressure (kPa)	Number	Volume	110 microns	210 micron
0°	200	83	250	7.9	35.9
	400	75	230	10.6	42.5
10°	200	97	258	5.0	31.2
	400	83	252	8.1	35.7
20°	200	133	269	2.4	26.0
	400	95	256	5.2	32.5
30°	200	204	291	0.5	15.9
	400	124	259	3.3	31.5
40°	200	227	306	0.2	10.7
	400	152	238	0.9	34.7

TABLE 3 -- Droplet Velocity profiles (m/sec) for SS8002 nozzle at 200kPa pressure

Position Angle	Distance (cm)	Mean of droplet size class (microns) 110	210	300	400	500
0°	20	6.3	9.9	11.4	12.4	15.3
	30	3.2	7.4	9.5	10.6	12.0
	40	2.3	5.5	7.8	9.6	11.2
	50	2.0	4.1	6.8	8.9	9.8
10°	20	5.6	9.9	11.4	12.8	12.6
	30	2.8	7.4	9.3	11.1	12.3
	40	2.0	5.3	8.0	9.6	10.4
	50	1.7	3.7	6.3	8.2	8.9
20°	20	5.1	9.7	11.1	12.3	14.1
	30	2.5	6.9	9.1	10.6	11.9
	40	1.7	4.7	7.6	9.2	11.0
	50	1.4	3.0	6.0	8.0	8.7
30°	20	4.5	9.6	10.9	12.5	14.9
	30	1.9	6.1	8.5	10.1	11.4
	40	1.8	4.2	7.1	9.1	10.8
	50	1.5	2.6	5.7	8.1	8.6
40°	20	2.6	8.1	10.3	12.1	13.4
	30	2.8	5.7	8.9	10.6	12.5
	40	2.2	3.4	6.9	9.1	10.2
	50	1.5	1.6	4.9	7.3	8.5

TABLE 4 -- The effect of transverse air flow on the droplet size distribution for a Hardi 4110-12 nozzle at 240 kPa, on the central axis

Distance from nozzle (cm)	Airflow (m/sec)	Median Diameter Number	Volume	Volume % less than 110 microns	210 micro s
20	0	66	238	11.8	38.5
	1	105	251	6.0	33.9
	2	139	260	2.7	29.8
30	0	70	230	13.2	41.7
	1	172	266	0.6	25.4
	2	207	272	0.1	19.4
40	0	76	244	9.9	37.8
	1	221	279	0.1	15.5
	2	250	301	0.0	6.1
50	0	82	228	8.4	41.6
	1	236	278	0.0	9.8
	2	272	311	0.0	2.2

Discussion

Droplet size data: The data obtained at 200kPa pressure is shown in Table 1. From this the following points are apparent; (1) The majority of the fine droplets are in the central core of the spray cloud and carried by the entrained air effect [11], (2) with increasing radial angle and distance from the nozzle the fine population decreases rapidly; for example at 30° and 40 cm the Number Median Diameter has increased to 204 microns and the population less than 110 microns has decreased to 0.5% by volume, (3) at a given angle the Volume Median Diameter is approximately constant irrespective of distance from the nozzle, even though the fine component is decreasing. The data in Table 2 illustrates the effect of increasing the applied pressure; at a distance of 40cm as representative. As would be expected there has been a significant increase in the fine component of the cloud.

Droplet velocity data: Table 3 shows a summary of the droplet velocities measured for five size classes, each of 15 microns width, throughout the spray cloud - at 200kPa pressure. From this it is clear that both axial and radial position have an effect on the measured droplet velocity.

For a given size class the velocity decreases with both distance from the fan and increasing radial angle. At a given position there is a characteristic droplet size/velocity profile as previously reported [8]. As might be expected an increase in spray pressure from 200kPa to 400kPa causes an overall increase in the velocity profiles throughout the spray cloud. From these droplet size and velocity profiles it is clear that the spatial population within the spray cloud is heterogeneous; and that single point measurements may not give meaningful data -depending on the intended use of the data. The data show that the bulk of the fine droplet component (less than 210 microns) has slowed down to a speed comparable with the forward speed during spraying by the typical target distance. In a travelling situation it is reasonable to assume that the entrained air effect will be disrupted and the fine droplets will have the potential to be significantly displaced from their original trajectories.

The removal of the fine droplets by lateral movement has been simulated for quantification purposes by directing a horizontal, transverse air-flow at the spray cloud during measurement with the PMS laser in a wind tunnel. Data is presented in Table 4 for a Hardi 4110-12 nozzle (equivalent to the SS8002) spraying at 240kPa pressure, measured on the central axis in still air, and transverse air flows of nominally 1.0 m/sec and 2.0m/sec. From this it is suggested that even an air flow of 1.0 m/sec is sufficient to displace approximately 50% of the droplets below 110 microns diameter at 20cm from the nozzle.

At greater distances virtually all these droplets are displaced. This suggests that the central entrained air plume is a highly metastable phenomenon that is effectively destroyed by even slight air flow or spray cloud movement. A spraying speed of 2 m/sec is very representative of many situations and it is clear that such a speed has a great effect on the spatial composition of the spray cloud. The above data is typical of many measurements and now gives support to the concept of motion induced droplet release from the primary spray cloud. Once released they become available for entrainment in the aerodynamically induced vortices created by pushing the spray cloud through the air during motion. As previously stated it is believed that very little air actually passes through the spray cloud, so the air must flow round it; in doing so the fine droplets are drawn away from the centre of the cloud. It is suggested that this results in the curling vortices observed and that these 'entrain' the fine droplets to form the characteristic trailing plumes [6, 7]. The magnitude of this effect will depend on the particular parameters used; but clearly one would expect the effect to be more pronounced for fine sprays travelling quickly.

CHARACTERIZATION OF THE TRAILING PLUME

The above data has indicated those droplets that appear to be at risk of being lost from the primary spray cloud. If that was the case it was presumed they became entrained in the trailing plume. It was therefore felt that an attempt to measure the droplet spectrum of the plume down-wind from a stationary nozzle in an airflow might support this concept.

Experimental Method

A single nozzle was used as previously described, and mounted 50cm above the canopy of a 1.2m x 1.5m grid of 30cm tall cotton plants in 7.5cm plastic pots. The plant canopy was used to give a realistic surface over which the airflow and spray cloud could interact. The PMS laser probe was mounted horizontally at the canopy level such that the laser beam was 40cm below the nozzle, 120cm downwind from the nozzle and 25cm from the centreline of the spray cloud. This was judged by observation to be the nominal centreline of one vortex and so a representative place to sample. Measurements were made for a range of Hardi 4110 series nozzles (-12 to -30) over the pressure range 150kPa to 500kPa in an air flow of nominally 2 m/sec. in a wind tunnel.

Results

The results in Table 5 show that the droplet spectrum of the trailing plume is essentially constant regardless of the operating conditions for this nozzle series, with a mean Volume Median Diameter of 114 microns.

TABLE 5 -- Volume Median Diameter of the trailing plume from a series of Hardi nozzles in an air flow of 2 m/sec

Nozzle Size	Pressure (kPa)	Flow rate (1/minute)	Volume Median Diameter (microns)
12	150	0.52	112
	200	0.60	120
	300	0.73	115
	400	0.85	110
	500	0.95	104
16	300	1.11	113
20	150	1.13	117
	300	1.57	117
	500	2.06	108
24	300	2.08	117
30	150	2.08	111
	300	2.97	116
	500	3.79	114

Discussion

This suggests that there must be some balance of forces resulting in the same droplet population being entrained in the vortices even though the primary spray cloud size distributions increase over the nozzle range. To say that the droplet population in the trailing plume is the same whether from a fine or coarse nozzle at low or high pressure is perhaps unexpected, but that is the conclusion from this data. This also suggests that during normal spraying a trailing plume with this size distribution (for this nozzle series) is laid out behind the primary spray cloud and is potentially available for further entrainment in any disturbed air behind the sprayer - or by ambient wind. Whether or not this trailing plume settles out onto the intended target behind the primary spray cloud, but in the spray path, is entirely dependent on these subsequent air currents. In the "still" air of a spray chamber this is indeed the case. In the outdoor situation there is always some off-target air movement (a 1m/sec wind is barely discernible without instrumentation) and there is bound to be some disturbed air flow from the wake of the sprayer itself.

Having developed a hypothesis for the formation of the trailing plume by entrainment into the induced vortices, and measured the characteristic droplet spectrum, the next question must be "what proportion of the spray emitted from the nozzle is lost from the primary spray cloud - or is potentially available to drift?" The other significant consequence of this hypothesis is that the origin of the driftable component is the trailing plume and not the primary spray cloud. This is an important difference because the droplet dynamics are very different. It suggests that any droplet that survives the effect of motion and remains within the primary spray cloud will not be available for off-target drifting. It is the finer, slower droplets that do not have sufficient energy to stay in the primary spray cloud that are drift susceptible. This may be an area overlooked by modellers, who may consider the droplet source as directly under the nozzle. In reality the situation may be much closer to an aerial spray swath where the droplet source may be considered as a line source behind the sprayer - with the droplets lacking any significant downward energy.

QUANTIFICATION OF THE TRAILING PLUME

The Drift Potential Concept

It was apparent during the course of this work that it would be more practical to attempt to quantify what reached the target area than what was lost in the trailing plume. Visual observation of the spray cloud in the air flow indicated that to all intents and purposes the deposited spray landed within a short distance behind the nozzle; and that by 1 metre downwind the droplets were travelling in the horizontal airstream. After initial experimentation the concept of measuring the Drift Potential was developed; where this term was defined as "that proportion of the spray output from the nozzle that fails to be deposited within a plan area of 1.05 metre width by 0.75 metre length at a standard distance below the nozzle and in an air flow of a defined speed".

Experimental Method

A matrix of 35 plastic, open topped, cubic containers of 15 cm x 15 cm cross section were arranged in 5 rows beneath the nozzle, such that the nozzle was directly above the centre container in the second row. All the containers were the same weight, within 0.5% tolerance. The tops of the thin-walled containers were clipped together such that a complete collection area was formed. The depth of the containers was sufficient to minimise any splashing and for the downwind containers the exposed inner walls were able to collect droplets with a shallow trajectory. The matrix was covered with a rigid sheet, the spray started at the required pressure and the airflow set. When the system was in equilibrium the sheet was quickly removed and the spray collected for a number of minutes, generally such that 2-3 litres were sprayed.

The containers were then recovered and the spray stopped.
Each container was then weighed to give a direct measure of the
volume of liquid collected. From this the total volume collected
was expressed as a percentage of the known total output from the
nozzle. Clearly from this the percentage lost was obtained. The
data can also be used to form a three dimensional patternator
concept.

Results

Typical results are shown in Table 6 for the series of Hardi
110° nozzles at 35cm target distance at 300kPa pressure. This
gives data for still air, and nominal 1m/sec and 2m/sec air
flows. From this it can be seen that the collection efficiency of
the technique is good (mean of 97.5% in still air). With
increasing air flow the percentage recovery decreases - the Drift
Potential increases - as might be expected. It also shows that
the Drift Potential decreases with increasing nozzle size.

TABLE 6 -- % loss - The Drift Potential - for a range of Hardi
110° nozzles in still air and air flows of 1.0 m/sec and
2.0 m/sec.

Nozzle Size	Still air	1.0 m/sec airflow	2.0 m/sec airflow
08	2.6	13.3	31.2
12	3.4	13.5	24.1
20	2.9	6.4	11.0
30	0.9	9.9	12.2

Discussion

Data has now been collected for a wide range of nozzle sizes
and makes and equivalent data obtained. Bearing in mind the
original intent of evaluating the loss in a real situation, most
data has been collected at 2m/sec. Three series of nozzles are
shown in Figure 1;
Spraying Systems 80° and 80° Low Pressure nozzles at 50 cm and
300kPa or 100kPa, and 110° nozzles at 35 cm and 300 kPa.

Two significant factors are clear :-

1.   The 80° nozzles at 50cm height and the 110° nozzles
     at 35cm height give equivalent results.

2.   The Low Pressure series gave significantly lower
     Drift Potential values than the normal pressure
     series.

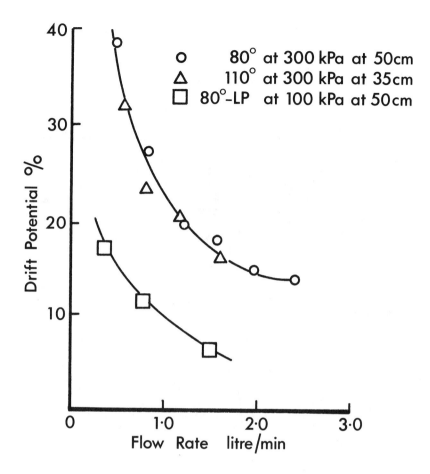

FIGURE 1 -- % loss - The Drift Potential - for Spraying
Systems nozzles in a 2.0 m/sec air flow

In this paper novel experimental data have been used to suggest an
explanation of the observed behaviour when spray clouds from fan
nozzles are moved or subjected to a headwind. This is important for
two reasons : (1) It helps to explain the fact that the deposition
observed from a single nozzle in a track sprayer does not necessarily
match its expected pattern profile. (2) It proposes an explanation
for the formation of the trailing plume observed behind a moving spray
cloud purely as a consequence of nozzle motion; and this lays out a
low energy droplet population that is then subject to drift.

It is important to stress at this stage that the concept of
the Drift Potential here is quantitative - but relative - and that
much of the spray within the plume may land within the adjacent
target area and not be "lost" to the surrounding area.

The values indicate the quantity actually landing under the sprayer; and show some agreement with other quoted values [12], where a loss of 25% was recorded for 8004 nozzles at 310 kPa spraying at 2.8 m/sec in the field. It is clear that the possible move to lower spray volumes and consequently finer sprays (and perhaps higher speeds) is likely to make the drift problem worse.

It is for this reason that increasing interest is being shown in equipment to force the spray cloud down into the crop - for example the Airtec nozzle and the Hardi Twin system, both of which use additional air to modify the droplet trajectories. However, the best way to reduce the drift problem may be to reduce the formation of the trailing plume - which means reducing the vortex effect. This is either an airflow problem for the aerodynamicist or a spray formation problem for the spray physicist - if there were no small droplets they could not become entrained in the vortices.

To support this work a unique series of tests have been carried out in conjunction with Hardi and Harper Adams Agricultural College. An 18m boom sprayer was used in a large closed barn with a bare soil surface. The sprayer was video filmed at night in still air driving at normal speeds away from the camera mounted on an elevated platform. Four shielded spotlights at the far end of the barn at ground level were used to illuminate the spray. This has shown that, for a wide range of nozzle sizes, pressures and travel speeds, each nozzle on the boom produces a characteristic pair of vortices behind the primary spray cloud. These are laid out very clearly for a considerable distance behind the boom - without mixing - until they are caught in the general air turbulence. This supports the belief that the effects produced in the spray chamber closely resemble the real situation.

CONCLUSION

The experimental work described here set out to quantify some of the factors believed to be responsible for the observed formation of vortices and a trailing plume behind a moving spray cloud. From this, three factors appear to be important - the initial formation, the movement of the spray and the release of the low energy droplets. It is clear that the appropriate choice of nozzle and operating conditions can be important in reducing this release. The concept of the Drift Potential is an attempt to quantify how much of the spray escapes from the immediate target area. The definition used here was developed to meet the initial requirements of the concept, and may require future refinement. It must be appreciated that the above data relates to water; formulated products and/or adjuvants are likely to change the behaviour to varying extents, as will the operating conditions.

ACKNOWLEDGEMENT

I am indebted to L Jorgensen of Hardi International, Denmark for collaboration with the original spray chamber video filming, and to W. A. Taylor of Hardi UK for the filming with the boom sprayer. The latter would not have been possible without the enthusiastic collaboration of staff at Harper Adams Agricultural College.

REFERENCES

[1]   Particle Measuring Systems, Inc., Boulder, Colorado, 80301,
[2]   Miller, P.C.H. and Hadfield, D.J., "A Simulation Model of the
        Spray Drift from Hydraulic Nozzles", Journal of
        Agricultural Engineering Research, Vol. 42, 1989, pp. 135-
        147.
[3]   Miller, P. C. H., AFRC. Silsoe, Bedford, UK. Personal
        communications.
[4]   Gohlich, H., 5th CIGR Congress. East Lansing, Michigan.
        1979, Paper P.1.5
[5]   Gohlich, H., "Assessment of Spray Drift on Sloping
        Vineyards", Crop Protection, Vol 2, No 1, 1983, pp 37-49
[6]   Anon, "Tackling the Problem of Drift by Adding Air,"
        Agriculture International, Vol. 40, No. 6, 1988 p. 141.
[7]   Cleanacres Machinery Ltd., Hazleton, Cheltenham,
        Gloucestershire, UK.
[8]   Young, B. W., "Practical Applications of the PMS 2-
        dimensional Imaging Spectrometer," Pesticide Formulations
        and Application Systems : Fifth Volume, ASTM STP 915, L.
        D. Spicer and T. M. Kaneko, Eds., American Society for
        Testing and Materials, Philadelphia, 1986, pp.128-133
[9]   Roth, L. O. and Porterfield, J. G., "A Photographic Spray
        Sampling Apparatus and Technique," Transactions of the ASAE
        Vol. 6, No. 4, 1965, pp.492-496.
[10] Bode, L. E., University of Illinois, Urbana. Personal
        communications.
[11] Goering, C. E., Bode, L. E. and Gebhardt, M. R.,
        "Mathematical Modelling of Spray Droplet Deceleration and
        Evaporation," Transactions of the ASAE, Vol. 15, No. 2,
        1972, pp.220-225.
[12] Clipsham, I. D., "The Influence of Target Area on the
        Variability of Spray Deposits," BCPC Monograph No 24 -
        Spraying Systems for the 1980's, J. O. Walker, Ed. 1980,
        pp.133-138.

Andrew J. Adams, Andrew C. Chapple and Franklin R. Hall

DROPLET SPECTRA FOR SOME AGRICULTURAL FAN NOZZLES, WITH
RESPECT TO DRIFT AND BIOLOGICAL EFFICIENCY

REFERENCE: Adams, A. J., Chapple, A. C. and Hall, F. R.,
"Droplet Spectra For Some Agricultural Fan Nozzles, With
Respect To Drift And Biological Efficiency." Pesticide
Formulations and Application Systems: 10th Volume, ASTM
STP 1078, L. E. Bode, J. L. Hazen and D. G. Chasin, Eds.,
American Society for Testing and Materials, Philadelphia,
1990.

ABSTRACT: The differences in drop size characteristics
along the long axis of "fine", "medium" and "coarse"
sprays produced by agricultural fan nozzles were determined using the Aerometrics Phase Doppler Particle Analyzer
(PDPA). Water, water + 0.06% adjuvant, water + 0.4% emulsifiable concentrate (EC) and water + 0.2% dispersible
granule (DG) were sprayed through each nozzle. There was
little difference in volume median diameter ($D_{v.5}$), number
median diameter ($D_{n.5}$) or percent (by number and volume)
of the spray contained in drops <100um or >300um diameter
when water was compared with the DG. However, the EC had
a significantly larger $D_{v.5}$ than the water sprayed through
five out of the six nozzles tested. Therefore, sprays
produced by specific nozzles cannot be reliably defined on
the basis of data for water alone. The percentage of
spray volume contained in drops <100um diameter (most
likely to drift) was 4 times greater in fine compared to
coarse sprays. However, the 6-fold difference in flow-
rate results in the higher throughput nozzles producing
most "driftable" volume per unit time. In coarse sprays,
69 - 84% by volume, but only 13 - 29% by number, was contained in drops >300um diameter (may rebound upon
impaction). When these drops were artificially removed
from the analysis, the $D_{v.5}$ was in the range 190 - 231um,
irrespective of nozzle or formulation. Thus, while coarse
sprays contain, characteristically, large drops (which may
contribute to off-target contamination because they are
not retained well by plant surfaces), they are not associated with a reduced drift potential compared to fine
sprays, based on volume production of drops <100um
diameter.

KEYWORDS: Agricultural sprays; drift; drop size; formulation; laser; nozzles; phase doppler particle analyzer.

Dr. Adams is a post-doctoral research associate, Mr.
Chapple is a graduate research associate and Professor Hall is
Head of the Laboratory for Pest Control Application Technology
(LPCAT), The Ohio Agricultural Research and Development Center,
Ohio State University, Wooster OH44691.

Characterization of agricultural sprays and determination of spray distribution patterns usually involves a measure of volume as the main index eg. the volume median diameter ($D_{v.5}$) / number median diameter ($D_{n.5}$) and the uniformity of spray volume collected beneath a boom in a static patternator. However, a single index cannot adequately describe the complex droplet spectra produced by a hydraulic nozzle [1] since it is possible to have numerous drop spectra, with identical $D_{v.5}$ and $D_{n.5}$ values. While many studies quote $D_{v.5}$ or application rate (by mass of active ingredient [AI] or carrier volume) or deposit per unit area, there are few examples where the actual deposit, described in terms of its prior drop size-/spectrum, spread, concentration, number per cm^2, distribution on target, etc., has been correlated with the biological effect There are even fewer studies where the required deposit has been defined prior to application. In addition, quantitative measures are invariably used to assess deposit distribution (eg. fluorometry) which may not provide an accurate measure of biological effectiveness. As an example, Cayley et al., [2] reviewed an extensive series of field trials comparing the performance of electrostatic with hydraulic sprayers and demonstrated that the increased deposits obtained using the former (at the same mass application rate) were not always reflected in the ultimate biological effect. Indeed, the distribution of the deposit, and the mode of action of the AI in relation to specific pests and diseases, were the key factors which determined the success or failure of an application.

Under laboratory conditions, uniformly-sized droplets may be applied directly onto the target surface. Consequently, the effects of spray deposit micro-distribution may be identified. Thus, spray coverage (% surface area) was shown to be an important factor determining the efficacy of dinocap against apple powdery mildew [3] such that, for equivalent dose/unit area, more drops of low concentration were more effective than fewer drops of higher concentration. However, the improved efficiency of utilization of AI observed when 175um droplets were compared with 400um droplets was not continued when droplet size was reduced further to 100um [3]. In contrast, Stevens and Bukovac [4] demonstrated that the percent uptake of diaminozide and 2-4-D-triethanolamine by field beans was not related to drop size, drops/cm2, leaf coverage, concentration of AI or application volume but was inversely related to the dose applied. Investigations using insecticides comprise the majority of the uniform droplet studies conducted to date, and these will be discussed in a subsequent section.

During the transport phase between atomization and deposit formation, droplets of different sizes will be affected to differing degrees by meteorological, physical and physico-chemical factors [5, 6]. The transport of small drops (<100um diameter), for example, will be dominated by air movement, which may be due to wind, turbulence caused by the sprayer, or the entrainment of the spray as it is directed towards the crop. Furthermore, this fraction of the spray will be most susceptible to evaporation. In theory, if droplets much smaller than 100um enter the canopy, they will almost certainly be retained upon impaction with plant surfaces, while the kinetic energy of larger droplets may exceed energy losses due to surface friction as they spread in which case they rebound instead of being retained [7]. However, high speed filming has demonstrated that droplets of water as small as 67um may rebound from a reflective leaf surface like broccoli [8], although the incorporation of surfactants may reduce or eliminate this phenomenon. Impaction velocity and drop trajectory will also be influential. Since the movement of drops >300um is dominated by gravitational forces [5, 7], it is highly unlikely that they will form

deposits on the undersides of leaves when crops are sprayed from
above. Consequently, in practice, these droplets may contribute
little to the control of many insect pests or fungal diseases unless
the AI exhibits translaminal or systemic activity. In cases where
total mass dose deposited is the most important deposit characteristic, large drops may only be retained upon second or subsequent
impact on foliage [8], and are more likely than smaller drops to be
deposited on the ground where they contribute to off-target contamination. Conversely, the movement of drops <150um becomes increasingly affected by air movement, leading to an increased possibility
of impaction on the undersides of leaves if the drops enter a canopy
where there is sufficient air movement, and also an increased likelihood of contributing to off-target waste as "drift" if they do not
[5].

INSECTICIDE DROPLET STUDIES.

Some investigators have used spinning discs to produce a narrow
droplet size spectrum, which allows the examination of the relative
effects of various spray parameters upon insecticide efficacy. In
this way, application rate, or dosage, was shown to be the most
important factor when lepidopterous pests were exposed to deposits
of *Bacillus thuringiensis* and *Baculovirus heliothis* [9]. This was
also shown to be true for aqueous permethrin sprays [10] applied
with hollow cone nozzles. Droplet size was as important as dosage
when the pyrethroid was applied in an oil-based carrier with smaller
drops providing better control with the same dose of AI [10].

Laboratory studies, in which droplets of uniform size have been
applied to show the effects of changing droplet size, concentration
of AI etc., are summarized in Tables 1 and 2. One noteworthy
feature is that the generation of droplet sizes below 50um is common
when oil-based formulations are in use (Table 2), but studies with
aqueous sprays have only recently examined droplets <100um. In many
cases, the emphasis was placed upon describing the most efficient
deposit (i.e. least AI/cm2) causing a median effect, and not upon
identifying the most important deposit parameter (eg. drop size,
drops/cm2, concentration of AI, etc.). The range in effective dose-
/cm2 illustrates the significance of deposit "quality" upon the
biological impact of any given dose. However, limits to this trend
were suggested when permethrin efficacy against *Plutella xylostella*
was investigated, since there appeared to be a minimum effective
dose/cm2 ([23], Table 2). A similar conclusion was reached for the
pickup of phosmet particles by *Laspeyresia pomonella* [24]. In what
is probably the most comprehensive study, an optimal concentration
of dicofol (11.8g/l) was derived for each of 5 droplet sizes,
although smaller droplets always utilized the AI more efficiently at
each concentration ([18], Table 2). Some authors evaluated the
influence of additional factors such as the effects of different
leaf surfaces upon spread, diffusion and pickup of AI ([20, 21],
Table 2), or changes in formulation and adjuvants ([21, 22], Table
2), or different viscosities of carrier oils [25]. It is evident
from the footnotes to each Table that there were considerable
differences in experimental design and interpretation of results.
However, almost all the studies indicated that the most efficient
utilization of AI was associated with the smallest droplet size and
the lowest concentration of the pesticide, although under field
conditions the efficiency of the deposit would have to be weighed
against the efficiency of obtaining that deposit.

Table 1. Laboratory studies of aqueous spray deposit parameters influencing efficacy of insecticide and miticides applied in uniformly sized drops.

Pest Species	Stage	Range Drop Diameter (μm)	Range Median Lethal Dose (ng/cm²)	Most Efficient Deposit	Reference
Panonychus ulmi	Adult	100-324	300-5000[a]	112μm; 2g/ℓ	[13]
	Adult	120-200	7-14[b]	120μm; 0.18 g/ℓ	[14]
Aphis gossypii	Adult	60-160	0.5-22.9[c]	120μm; 0.0075g/ℓ	[15]
Lymantria dispar	2nd instar	50-350	44.9-137.1[d]	100μm; 4227IU/ℓ	[16]
Trichoplusia ni	1st instar	60-120	0.3-15.8	60μm; 0.0076g/ℓ	[17]

[a] Dose-response data derived from a single drop density for each size/concentration combination.

[b] Dose causing 80% reduction in egg deposition.

[c] Dose eliciting hyperactivity.

[d] Dose in International Units (IU) of B.t. Data estimates based on numerical mid-point of drop size classes 50-150, 150-250 and 250-350 μm diameter.

Table 2. Laboratory studies of oil-based spray deposit parameters influencing efficacy of insecticides and miticides applied in uniformly-sized drops.

Pest Species	Stage	Range		Most Efficient Deposit	Reference
		Drop Diameter (μm)	Median Lethal Dose (ng/cm^2)		
Tetranychus urticae	Eggs	18-100	12-86	18μm; 1.18g/ℓ	[18]
	Larvae	18-100	5-22	18μm; 1.18g/ℓ	[19]
	Protonymphs	18-100	6-138	18μm; 1.18g/ℓ	
Trialeurodes vaporariorum	1st instar	59-120	125-6310	59μm; 5g/ℓ [a]	[20]
	1st instar	31-108	80-1650	31μm; 10g/ℓ [a]	[21]
	1st instar	39-47	42-442	40μm; 10g/ℓ [b]	[22]
Plutella xylostella	2nd instar	42-216	7-60	50μm; 5g/ℓ	[23]

[a] Concentration and drop size effects assessed independently.

[b] Investigated effects of formulation and surfactants in spray.

OBJECTIVES

In addition to the biological result, agricultural sprays may cause, or (perhaps more significantly) be perceived to cause, an environmental hazard as "drift" or through contamination of groundwater. Thus, Burt and Smith [11] concluded that where drift was a potential problem, drops <140um should not be applied, while recognizing that effective and efficient pest control may require those smaller droplets. In many respects, this contradiction encapsulates the dilemma we shall address in this paper: how can a spray satisfy biological and environmental requirements? The laboratory studies reviewed above examined droplet deposit parameters and their influence upon biological efficacy and suggest that the optimum droplet size for most efficient use of AI is <120um for water based spray and <60um for oil based sprays. Yet this is the component of sprays that constitutes the potential for "drift". The work described below compares the characteristics of "fine", "medium" and "coarse" sprays produced by flat-fan nozzles, as defined by the British Crop Protection Council in 1985 [12]. This description classifies nozzles by manufacturer into categories of fine, medium, etc., but does not reference droplet spectra. For this work, this was considered a reasonable starting point, as the user is more likely to think in qualitative rather than quantitative terms.

The droplet spectra were determined for water and various 'blank' formulations. By examining spray drop spectra, we shall consider whether larger orifice nozzles provide a reliable means of reducing the "driftable" (drops <100um diameter) component of a spray and, if so, at what cost, since one 300um drop deposited on the ground is as much off-target pesticide as 27 100um drops carried away in the air. It is not our intention to provide recommendations, or to provide a comprehensive description of the spray produced by every type of nozzle, across its range of operating pressures, for every adjuvant/pesticide concentration combination etc., that might be sprayed through each nozzle, since that is clearly a task well beyond the scope of a single paper. However, it is hoped that the data presented will help to emphasize the need for such a database.

MATERIALS AND METHODS

Nozzles and Spray Liquids.

Three 80° fan nozzles from two manufacturers (Spraying Systems, Wheaton, IL and Hartvig Jensen & Co. A/S, Glostrup, Denmark) were examined. These nozzles were selected with reference to the BCPC Nozzle Selection Handbook [12] for their production of "fine" (XR8001VS, 2080-10), "medium" (XR8003VS, 2080-16) and "coarse" (XR8006VS, 2080-30) sprays. It should be noted that these nozzles cannot be directly compared: for example the 2080-30 is more closely equivalent to an XR8008VS. Also, while in practice fan nozzles are used to apply many pesticides, they are not generally recommended for use with fungicides and insecticides, hollow cones being preferred. In each case, spray characteristics were determined using tap water, water + 0.06% adjuvant Ortho X-77 (Chevron Chemical Co., Richmond CA), water containing 0.4% of the blank emulsifiable concentrate (EC) formulation of "Pydrin 2.4 EC" (Du Pont, Wilmington DE), and water containing 0.2% of a dispersible granule (DG) formulation ("Manzate 200 DF"; Du Pont, Wilmington DE). The concentration of Ortho X-77 represents the upper and lower ends of the recommended rate for insecticides/fungicides and herbicides respectively. The concentrations of the EC and DG were chosen as representative of pesticide sprays applied in the field.

Spray Characterization.

A Phase Doppler Particle Analyzer (PDPA, Aerometrics Inc.,
Sunnyvale CA), calibrated at Aerometric's laboratories, was used to
measure the droplet size and velocity distribution of each spray.
At present, there is very little published data generated by the
PDPA for agricultural sprays, and none of that includes measurements
across the same expanse of spray that is included in data sets

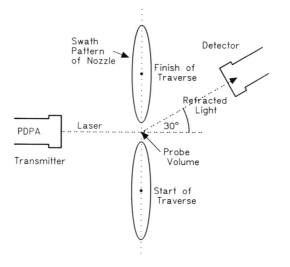

FIGURE 1. View from above to show position of PDPA and spray
          being characterized.

available from a Malvern Particle Sizer. Since there is no accepted
standard for this kind of spray characterization with the PDPA, we
feel it is important to be as thorough as possible in our description of the operating parameters of the equipment used in this
study. Calibration of the PDPA was set using the XR8001VS spraying
water and measuring the drops in the center of the spray pattern.
The set-up calibration varied by ±5um (3.8%) over the period of two
weeks in which the tests reported here were conducted. The operating and measurement parameters for each test are given in Table 3
and additional tests demonstrated that for coarse sprays, <0.01% of
the spray volume was contained in drops smaller than 26um. As
expected, there was approximately a 6-fold increase in flow rate
from the fine to the coarse nozzle of both manufacturers. Flow rate
was unaffected by the nature of the liquid being sprayed. The PDPA
operates over a 35-fold range in drop diameter and the ranges for
each nozzle/liquid combination were selected to maximize the validation of drops that the computer attempted to characterize.
Although the same lens combination was used for each measurement,
the diameter range was determined by selecting an appropriate maximum figure in the software. Since measurements are only made at a
single point, where the two beams intersect (the "probe volume") and
not along the length of a single beam, an x, y, z positioner ("Uni-

slide" Velmex Inc., E. Bloomfield NY) was programmed to move the
nozzle over the probe volume at a traversing speed of 0.25cm/sec
(Figure 1) at a vertical distance of 30cm. Thus, data was acquired
along the center of the long axis of the flat fan pattern over a
period of 280 seconds, which compares exactly with the measurement
made along the long axis of the spray using the Malvern (see below).

Table 3 -- Operating parameters of PDPA[a] and flow rates for
each nozzle.

Nozzle[b]	Liquid							
	Water		Orto X-77		Blank EC		DG	
	ml/min	um	ml/min	um	ml/min	um	ml/min	um
XR8001VS	347	11-400	341	13-450	350	13-450	360	11-400
2080-10	375	13-450	361	13-450	377	13-450	380	11-400
XR8003VS	1023	17-600	1010	19-650	1017	20-700	1010	20-700
2080-16	1077	17-600	1061	19-650	1067	20-700	1087	20-700
XR8006VS	1975	26-900	1943	26-900	1990	26-900	1967	26-900
2080-30	2423	26-900	2353	26-900	2420	26-900	2393	26-900

[a] In every case, the combination of 2000 mm transmitting and
160 mm collimating lens was used; photomultiplier tube voltage
was set at 325 volts.
[b] All nozzles operated at 207 kPa pressure.

Preliminary work with the PDPA has demonstrated that variations
in spray spectra measurement from run to run can be large if there
are <5000 droplets in the sample. Table 4 shows the effect of
number of droplets counted upon the $D_{v.5}$ for the identical fan and
operating conditions, and reproducibility of results (95% confidence
interval) for one of the calibration nozzles in LPCAT: a Spraying
Systems TeeJet 8001E operating at 207 kPa, 50cm above the probe
volume and with the nozzle fixed in position such that only drops in
the center of the spray were characterized. Each measurement in the
present study included over 12,000 droplets and the error was small.

Table 4 -- Effect of sample size upon $D_{v.5}$ and $D_{n.5}$ as
measured at the center of a water spray[a]
produced by a TeeJet 8001E nozzle (4 reps
per sample size).

Actual Counts:	Corrected Counts:	$D_{v.5}$ (±95% CI[b], um)	$D_{n.5}$
500	1041	194.5 (±82.8)	26.5 (±3.2)
2000	3110	188.2 (±20.0)	37.9 (±3.1)
5000	7819	196.2 (±8.7)	38.2 (±2.7)

[a] Operating parameters: 207 kPa; nozzle to probe volume 50 cm;
500 mm transmission and 300 mm collimating lenses; diameter
range 13 - 470 um.
[b] 95% Confidence Interval.

In the tables, the mean of 3 repetitions per nozzle/liquid combination is given. In addition to the $D_{v.5}$ and $D_{n.5}$ statistics, the percentage of drops (by number and volume) below 100um is presented as a measure of both "drift potential" and "biological usefulness". Similarly, the volume and number contributions of drops >300um (which we have defined as those drops with the greatest "bounce potential" and hence a low likelihood of forming deposits upon first impaction?) were identified. These arbitrary figures and the alternative descriptions are clearly open to debate and will be subject to numerous factors including wind speed, drop velocity, crop canopy structure, plant surface etc. as well as the mode of action of the pesticide being applied, and especially the inclusion of surfactants, 'stickers', etc. Finally, the data collected were reanalyzed after all the droplets >300um diameter were removed from the data sets to derive the resulting change in $D_{v.5}$ and $D_{n.5}$.

RESULTS AND DISCUSSION

Table 5 shows that the $D_{v.5}$ and $D_{n.5}$ for each nozzle were altered by the nature of the liquid being atomized. The $D_{v.5}$ and $D_{n.5}$ with the EC were always the largest and the $D_{v.5}$ was significantly higher than for water alone in all but one nozzle (Hardi 8020-30). These effects were paralleled (but not significantly) by the larger $D_{v.5}$ values for the adjuvant sprays, when compared with water. Further tests with higher concentrations of X-77 (up to 0.5%, which is the highest rate recommended for this product) showed no significant change in $D_{v.5}$ compared to water alone. There was little difference between the $D_{v.5}$ of water and DG sprays (which contained no surfactant) through each nozzle. Hence, a check of the differences between water and EC sprays was performed using an alternative system that does not rely on light passing through the drops. This was to ensure that the large differences recorded with the PDPA were real, and not an artifact due to emulsion inclusions. As can be seen from Table 5, % validations (ie. the number of signals received by the PDPA that are considered to be from droplets) had fallen from 90%+ for water, Ortho and the WG to c.85% for the EC. (Aerometrics suggest that 80% validation is perfectly acceptable.) It was thought that the PDPA was rejecting signals (up to 15%) because of the inclusion of minute oil droplets in the droplets of carrier water - ie. the emulsion. Hence, water and EC sprays were characterized with an XR8003VS nozzle, using a Malvern 2600/3600 Particle Sizer (data courtesy, Parker Hannifin Corp., Cleveland OH). The $D_{v.5}$ values were 248um and 308um, respectively, confirming that the EC was associated with a substantial increase in drop size. Significant effects of formulations upon drop size have also been shown for rice herbicide mixtures [26], while Krueger and Reichard [27] have demonstrated considerable changes in spray distribution across the swath for identical nozzles when emulsions were compared with water. Table 5 indicates that the PDPA gave figures of 325um and 368um which correspond with the Malvern data (see above). The substantial differences between the two measurement systems may be attributed to differences between temporal (PDPA) and spatial (Malvern) sampling (see [28, 29]).

The problem of reduced validations was exacerbated in other tests when the concentration of the EC was increased, and supports the idea that included particles interfere with the detection and measurement of refracted light (although the refractive index of the 4 liquids was essentially the same). On-going studies include the investigation of this phenomenon.

The small droplet (<100um diameter) component of each spray is shown in Table 6.   Including 0.4% EC in the spray liquid always resulted in the lowest volume of sub-100um drops. However, one interesting feature of the EC and DG sprays was that, in both cases, there was a 4-fold decrease in the volume of small drops from fine (XR8001VS) to coarse (2080-30)sprays, while the proportion by number remained at around 20% (EC) or 38% (DG). Since there was a

Table 5 -- Spray sample size and median droplet diameters for each nozzle with each spray liquid.

Spray Liquid		XR8001VS	2080-10	XR8003VS	2080-16	XR8006VS	2080-30
					Nozzle:		
Water	Counts	24744	21098	26181	26017	34144	28309
	(%Valid)	(97%)	(96%)	(93%)	(93%)	(92%)	(91%)
	$D_{v.5}$	199	224	325	313	390	454
	$D_{n.5}$	103	112	126	125	124	127
Ortho	Counts	23614	22079	21247	21328	23971	28106
X-77	(%Valid)	(96%)	(95%)	(90%)	(90%)	(92%)	(89%)
	$D_{v.5}$	217	237	357	361	404	459
	$D_{n.5}$	111	118	180	169	134	145
Blank	Counts	16759	14459	20165	22040	14773	12514
EC	(%Valid)	(85%)	(87%)	(89%)	(88%)	(86%)	(88%)
	$D_{v.5}$	229	248	368	376	422	457
	$D_{n.5}$	141	150	187	195	194	199
DG	Counts	17174	17321	23117	20975	24480	23189
	(%Valid)	(92%)	(90%)	(91%)	(93%)	(95%)	(93%)
	$D_{v.5}$	207	221	341	315	384	423
	$D_{n.5}$	113	119	124	116	128	128
L.S.D[1].:	$D_{v.5}$	5.7	5.4	7.9	13.6	8.7	10.7
	$D_{n.5}$	7.4	7.9	9.1	16.0	8.9	17.0

[1] LSD: P = 0.01

6-fold increase in flow-rate (Table 3), these data indicate that the absolute volume of spray produced per unit time that was contained in "driftable" drops was actually greatest in the coarse sprays. Clearly, any increase in travel speed when coarse rather than fine nozzles are used will result in a corresponding decrease in time to spray a given area, with a consequent reduction in the volume of spray contained in small drops. However, the reduction in "drift potential" may be offset

Table 6 -- Percentage of droplets < 100 um diameter in sprays for each nozzle for each spray liquid.

	% Volume				% Number			
	Water	Ortho X-77	Blank EC	DG	Water	Ortho X-77	Blank EC	DG
XR8001VS	9.3	5.8	2.5	7.0	48.9	41.9	23.8	42.0
2080-10	5.3	4.1	1.8	5.4	41.0	38.1	21.5	38.7
XR8003VS	2.0	0.9	0.6	2.0	34.8	19.9	17.3	35.6
2080-16	2.0	1.3	0.6	2.7	34.0	26.5	16.8	38.9
XR8006VS	2.1	1.7	0.7	2.1	38.0	34.6	19.6	35.6
2080-30	1.4	1.1	0.6	1.6	37.0	31.6	20.6	35.7

to some extent by the increase in turbulence and boom movement that
accompany higher speeds. It should also be noted that a "coarse"
nozzle is more likely to be used to apply a more dilute solution
than a "fine" nozzle; hence, "driftable" droplets will contain less
AI. Despite this, the point is made that an increase in nozzle
orifice size alone does not necessarily preclude a reduction in
drift.

Large drops (>300um diameter) comprised 70% or more of the spray
volume for all coarse sprays and only 10 - 30% of fine sprays, with
the EC providing the largest volume and number component in every
case (Table 7). However, the included surfactant might offset any
potential losses due to large drops rebounding from foliage by increasing the likelihood of these drops being retained upon impaction
[7]. When these drops were removed from the data sets, the derived
$D_{v.5}$ values hardly differed between formulations and nozzles
(Table 8). The corresponding $D_{n.5}$ values gave more variable
results. These data indicate that the primary difference between
coarse and fine sprays is the presence of many large (>300um) drops
in the former and their relatively low occurrence in the latter -
and not a substantial change in the volume of applied spray that may
be considered as a

Table 7 -- Percentage of droplets > 300 um diameter in sprays for
each nozzle for each spray liquid.

	% Volume				% Number			
		Ortho	Blank			Ortho	Blank	
	Water	X-77	EC	DG	Water	X-77	EC	DG
XR8001VS	9.9	15.2	20.3	10.7	0.9	1.7	3.3	1.1
2080-10	17.0	23.4	27.9	14.3	2.0	3.1	5.3	1.8
XR8003VS	57.5	67.1	69.4	60.6	10.1	19.2	21.0	11.0
2080-16	55.1	66.0	71.7	53.2	9.7	16.0	22.6	8.1
XR8006VS	71.5	73.0	78.2	69.4	13.1	15.1	25.7	13.1
2080-30	80.2	81.7	83.6	76.2	17.4	20.5	28.8	15.5

potential drift hazard. Thus, if it were shown experimentally that
less, in volume terms, of a coarse spray can be captured as drift
compared to a fine spray, for identical travel speeds, then, again,
a single numeric - 100um - value for a complex phenomena such as
drift may prove to be too inflexible. For example, if the large
drop fraction entrains the small drops during the transport phase
from atomization to deposition, then the small drops of a coarse

Table 8 -- Derived median droplet diameters if drops > 300 um are
ignored in every spray.

	$D_{v.5}$				$D_{n.5}$			
		Ortho	Blank			Ortho	Blank	
	Water	X-77	EC	DG	Water	X-77	EC	DG
XR8001VS	190	202	211	195	103	111	137	113
2080-10	205	210	220	207	111	113	145	118
XR8003VS	216	224	231	219	117	155	164	120
2080-16	216	221	232	216	119	140	161	112
XR8006VS	211	214	229	213	113	116	156	117
2080-30	213	220	229	214	119	120	148	123

spray would be more likely to enter a canopy or reach the target
compared to a fine spray. Under these circumstances, a figure
representing the "drift potential" in terms of spray below a certain
drop diameter would be inappropriate. Rogers ([30] and references
therein) has done trials with shrouded sprayers using low volume,
small droplet ($D_{v\ 5}$: 130um) herbicide sprays and claims negligible
drift and enhanced biological activity, compared with conventional
sprays ($D_{v\ 5}$: 410um). Supporting conclusions are also drawn in
another study described in this volume [31]. These data suggest
that coverage and distribution may be maximized with small droplet
sprays without introducing a drift hazard. It is possible that
unless total dose deposited is of sole importance (which does not
appear to be the case for insecticides), the "biological work" is
performed by the driftable drops that enter the canopy, while the
large drops are more useful as small drop transporters, via entrainment, than as chemical carriers. This emphasizes the importance of
understanding and appreciating drop transport (including drop velocity and trajectory, turbulence, entrainment etc.) as well as drop
spectra and distribution across and between nozzle swaths.

ACKNOWLEDGMENTS

We thank Roger Downer for his assistance with data collection
using the PDPA and for useful discussions, and Bob Fox, who wrote
the programming for the x, y, z plotter. Thanks also to Gary Bittel
of the Parker Hannifin Corp. who performed the tests with the
Malvern 2600/3600. We are also grateful to Joe Reed, D. L. Reichard
and R. E. Treece for their comments on the original manuscript.
Salaries and research support provided by State and Federal funds
appropriated to the Ohio Agricultural Research and Development
Center, The Ohio State University. Manuscript Number 88-90.

REFERENCES

[1]   Southcombe, E. S. E. The BCPC nozzle selection system.
       Proceedings ANPP International Symposium on Pesticide Applic-
       ation, Paris, France. 1988, pp 71-78.

[2]   Cayley, G. C., P. E. Etheridge, R. E. Goodchild, D. C.
       Griffiths, P. J. Hulme, R. J. Lethwaite, B. J. Pye and G. C.
       Scott. Review of the relationship between chemical deposits
       achieved with electrostatically charged rotary atomisers and
       their biological effects. British Crop Protection Council Mono-
       graph 28, 1985, pp 87-96.

[3]   Frick, E. L. The effect of volume, drop size and concentration,
       and their interaction, on the control of apple powdery mildew by
       dinocap. British Crop Protection Council Monograph 2, 1970, pp
       23-33.

[4]   Stevens, P. J. G. and Bukovac, M. J. Effects of spray application
       parameters on foliar uptake and translocation of daminozide and
       2-4-D-triethanolamine in *Vicia faba*. Crop Protection Vol. 6,
       1987, pp 163-170.

[5]   Johnstone, D. R. Physics and meteorology. In: Pesticide Applic-
       ation: Principles and Practice. Ed. P.T. Haskell 1985, pp 35-67.

[6]   Young, B. W. Studies on the retention and deposit characteristics
       of pesticide sprays on foliage. Paper P-1-4: Application meeting,
       IXth CIGR Congress, East Lansing, MI. 1979.

[7] Spillman, J. J. Spray impaction, retention and adhesion: an
     introduction to basic characteristics. Pesticide Science Vol. 15,
     1984, pp 97-106.

[8] Reichard, D. L. Drop formation and impaction on the plant. Weed
     Technology Vol. 2, 1988, pp 82-87.

9]   Smith, D. B., Hostfetter, D. L. and Ignoffo, C. M. Laboratory
     performance specifications for a bacterial (*Bacillus thuring-
     iensis*) and a viral (*Baculovirus heliothis*) insecticide. Journal
     of Economic Entomology Vol. 70, 1977, pp 437-441.

[10] Wofford, J. T., Luttrell, R. G. and Smith, D. B. Relative effects
     of dosage, droplet size, deposit density and droplet concentr-
     ation on mortality of *Heliothis virescens* (Lepidoptera:
     Noctuidae) larvae treated with vegetable-oil and water sprays
     containing permethrin. Journal of Economic Entomology Vol. 80,
     1987, pp 460-464.

[11] Burt, E. C. and Smith, D. B. Effect of droplet sizes on
     deposition of ULV spray. Journal of Economic Entomology Vol. 67,
     1974, pp 751-754.

[12] Anon. Nozzle selection handbook. British Crop Protection Council,
     Croydon, UK, 1985.

[13] Fisher, R. W., Menzies, D. R., Herne, D. C. and Chiba, M. Para-
     meters of dicofol spray deposit in relation to mortality of
     European red mite. Journal of Economic Entomology Vol. 67, 1974
     pp 124-126.

[14] Alm, S. R., Reichard, D. L. and Hall, F. R. Effects of spray drop
     size and distribution of drops containing bifenthrin on *Tetrany-
     chus urticae* (Acari: Tetranychidae). Journal of Economic Entomol-
     ogy Vol. 80, 1987, pp 517-520.

[15] Adams, A. J. and Hall, F. R. Initial behavioural responses of
     Aphis gossypii to defined deposits of bifenthrin on chrysanth-
     emums. Crop Protection Vol. 9, 1990, (in press).

[16] Bryant, J. E. and Yendol, W. G. Evaluation of influence of
     droplet size and density of *Bacillus thuringiensis* against Gypsy
     moth larvae (Lepidoptera: Lymantridae). Journal of Economic
     Entomology Vol. 81, 1988, pp 130-134.

[17] Adams, A. J. and Hall, F. R. Influence of bifenthrin spray
     deposit quality on the mortality of *Trichoplusia ni* (Lepidoptera:
     Noctuidae). Crop Protection Vol. 8, 1989, pp 206-211.

[18] Munthali, D. C. Biological efficiency of small dicofol droplets
     against *Tetranychus urticae* (Koch) eggs, larvae and protonymphs.
     Crop Protection Vol. 3, 1984, pp 327-334.

[19] Munthali, D. C. and Wyatt, I. J. Factors affecting the biological
     efficiency of small pesticide droplets against *Tetranychus
     urticae* eggs. Pesticide Science Vol. 17, 1986, pp 155-164.

[20] Abdalla, M. R. A biological study of the spread of pesticides
     from small droplets. PhD thesis, University of London, 1984.

[21] Adams, A. J., Abdalla, M. R., Wyatt, I. J. and Palmer, A. The relative influence of the factors which determine the spray droplet density required to control the glasshouse whitefly, *Trialeurodes vaporariorum.* Aspects of Applied Biology Vol. 14, 1987, pp 257-266.

[22] Adams, A. J., Fenlon, J. S. and Palmer, A. Improving the biological efficacy of small droplets of permethrin by the addition of silicon-based surfactants. Annals of Applied Biology, Vol. 112, 1988, pp 19-31.

[23] Omar, D. and Matthews, G. A. Biological efficiency of spray droplets of permethrin ulv against the diamondback moth. Aspects of Applied Biology Vol. 14, 1987, pp 173-180.

[24] Fisher, R. W. and Menzies, D. R. Pickup of phosmet wettable powder by codling moth larvae (*Laspeyresia pomonella* [Lepidoptera: Olethreutidae]) and toxicity responses of larvae to spray deposits. Canadian Entomologist Vol. 111, 1979, pp 219-233.

[25] Crease, G. J., Ford, M. G. and Salt, D. W. The use of high viscosity carrier oils to enhance the insecticidal efficacy of ULV formulations of cypermethrin. Aspects of Applied Biology Vol. 14, 1987, pp 307-322.

[26] Bouse, L. F., Kirk, I. W. and Bode, L. E. Effect of spray mixture on droplet size. Paper 891006 ASAE/CSAE Meeting, Quebec, Canada, 1989.

[27] Krueger, H. R. and Reichard, D. L. Effect of formulation and pressure on spray distribution across the swath with hydraulic nozzles. Pesticide Formulations and Application Systems Vol 4 ASTM STP 875, T. M. Kaneko and L. D. Spicer, Eds., American Society for Testing and Materials, Phila delphia PA, 1985, pp 113-121.

[28] Frost, A. R. and Lake, J. A. The significance of drop velocity to the determination of drop size distributions of agricultural sprays. Journal of Agricultural Engineering Research Vol. 26, 1981, pp 367-370.

[29] Young B. W. and Bachalo W. D. The direct comparison of three 'in-flight' droplet sizing techniques for pesticide spray research. 1988; in 'Optical Particle Sizing: Theory and Practice' pp 483-497 Eds. G. Govesbet and G. Grehan Plenum Press, NY.

[30] Rogers, R. B. Shrouded sprayer: Advantages in field application systems. Pesticide Formulations and Application Systems Vol 8 ASTM STP 980, D. A. Hovde and G. B. Beestman, Eds., American Society for Testing and Materials, Philadelphia, PA, 1989.

[31] Hall F. R. Reed J. R., Reichard D. L., Reidel R. M. and Lehtinen J. Pesticide delivery systems: spray distribution and partitioning in canopies. Pesticide Formulations and Application Systems: 10th Volume. ASTM STP 1078, L. E. Bode, J. L. Hazen and D. G. Chasin, Eds., American Society for Testing and Materials, Philadelphia, 1990.

Norman B. Akesson and Richard E. Gibbs

PESTICIDE DROP SIZE AS A FUNCTION OF SPRAY ATOMIZERS AND LIQUID FORMULATIONS

REFERENCE: Akesson, N. B. and Gibbs, R. E., "Pesticide Drop Size as a Function of Spray Atomizers and Liquid Formulations," Pesticide Formulations and Application Systems: 10th Volume, ASTM STP 1078, L. E. Bode, J. L. Hazen and D. G. Chasin, Eds., American Society for Testing and Materials, Philadelphia, 1990.

ABSTRACT: The use of the ruby laser automatic, in situ spray drop size and frequency analyzer has permitted a much more detailed, rapid and reproducible study of spray drop atomization than hitherto possible. With the advent of these machines and their use in pesticide spray studies, it is now possible to enumerate specific size and size range for the many different atomizers, fan, cone, jet hydraulic pressure types and rotary screen devices used for pesticide spraying. A given atomizer will respond to formulation changes as these affect surface tension, viscosity and density and will produce drop size in relation to these physical properties.

As a result of several years collection of field drift-loss data we have put together a model program which utilizes the drop size information as noted, combines this with weather factors and operational parameters of the spray aircraft to enable prediction of amounts of pesticide to be expected as fallout and airborne portion at specified distances downwind. This information will be of significant value to pesticide manufacturers, users and to registration and regulatory agencies charged with safe use of pesticide chemicals.

KEYWORDS: spray drops, liquid atomizers, liquid formulations, pesticide sprays, spray drift-loss, plant coverage, drift modeling

The use of toxic chemicals and biological agents as pest control materials for plant protection and disease vector control is a widely accepted and essentially desirable procedure needed by growers throughout the world to insure high quality produce and high productivity for profitable farming operations [1]. Naturally grown produce, implying no use of pesticides or commercial fertilizers, has received considerable news print in recent years, but there is little possibility for growers to be able to feed the world population without the use of these much maligned but essential plant protection and nutrition materials.

This is not to say that such things as integrated pest management and minimum use of chemicals should not be pursued. Certainly growers and consumers share a common goal to reduce use of chemical pesticides through more careful monitoring and use of pest resistant crop types as well as biological means such as natural predacious agents.

Norman B. Akesson is professor emeritus, University of California, Agricultural Engineering Department, Davis, California, 95616, and Richard E. Gibbs is a project engineer, Boeing Aircraft Company, Seattle, Washington.

But, for the present, growers are dependent on use of pesticides and we, as agricultural engineers, need to continue development of better, more accurate and less contaminating means for application [2]. This can be examined beginning at the atomization process, which, in spite of considerable study and research, remains a very inefficient system, generally producing a wide spectrum of drop sizes dependent on (a) type of atomizer such as hydraulic pressure, air shear, 2-fluid and various rotary devices, (b) formulations and adjuvants ranging from the most common of the surface tension agents to complex polymer materials and particulates, such as wettable powders and flowables [3]. Considerable progress has been made in defining the role of the atomizers in pesticide spray application [4, 8], but practical field use nozzles have not been greatly improved in their many years of use in agriculture. Essentially, we need to reduce the spectrum width for the atomized spray liquid. This can be accomplished in a laboratory setting through use of various monodisperse devices, but transferring this technique to field usage has not, so far, been accomplished.

Reviewing the accomplishments in atomization research for agriculture use reveals that there is promise in spectrum width reduction with (a) certain rotary types [5], (b) aerodynamic types for release into an airstream [6], and (c) air shear types also suited to airstream use, such as aircraft and air-carrier spraying.

An examination of these atomizers along with the effects of formulations on their capabilities indicates that there are definite patterns of drop size produced that are amenable to such techniques as regression analyses. With a knowledge of drop size parameters, we can couple this information to the field drift-loss data collected on deposit, coverage and drift-loss studies from various drop size sprays and obtain the basic data required for predicting the deposit on target crops and the losses downwind from specific spray applications applied under known conditions. This information is presently being developed into a computer program which can be coupled with other programs presently used for determination of spray application and calibration for volume and active chemical being applied. This is aimed at the operational level for use by commercial aircraft applicators, as well as by growers and enforcement agencies [7].

DROP SIZE ANALYSIS

Particle Measuring Systems Ruby Laser Analyzer

Figure 1 shows the wind tunnel facility at the University of California, Davis, used to study pesticide spray drop size in simulation of aircraft or air carrier ground sprayers. This uses a basic Particle Measuring Systems Company OAP-2D-GA-1, 2-dimensional

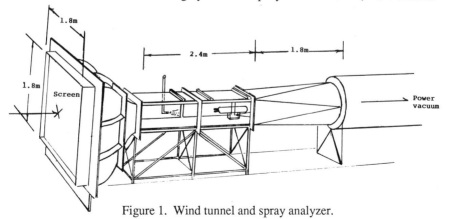

Figure 1. Wind tunnel and spray analyzer.

analyzer which categorizes the spray being tested into 64 channels having a size range from 28 μm to 2062 μm drop diameter. The probe is shown located in the wind tunnel throat, a 60 x 60 cm size, where air velocity can be varied from about 32 to 240 km/hr. The nozzle under test is traversed in 2 dimensions across the probe, thus giving a scan of the spray as it leaves the test nozzle. Further information on validation of this scanner is noted in reference [9].

Nozzles: The most widely used nozzles or spray atomizers are the hydraulic pressure types shown in Figure 2. From left to right these are: (a) tube jet, (b) orifice jet, (c) hollow cone (disc-core type), (d) hollow cone tangential type, (e) solid cone, (f) fan, and (g) deflector fan. Variations of these have been available for many years from nozzle manufacturing companies and with the exception of the fan type (f) were all designed for uses other than pesticide application. Figures following show the drop size characteristics of these nozzles, all produce a wide range of sizes which can be moved upward or downward to produce different average or median sizes, basically by increasing the nozzle size and flow rate. An increase in pressure usually results in smaller drop size even though flow may also be increased in this manner.

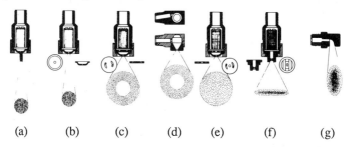

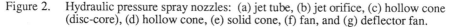

    (a)   (b)   (c)   (d)   (e)   (f)   (g)

Figure 2.   Hydraulic pressure spray nozzles: (a) jet tube, (b) jet orifice, (c) hollow cone (disc-core), (d) hollow cone, (e) solid cone, (f) fan, and (g) deflector fan.

Figure 3 shows a typical distribution pattern for an orifice jet nozzle, 0.24 cm (6/64 in) dia. operated in a 161 km/hr airstream at 0° or discharged with the air at 276 kPa liquid pressure. As can be seen, the spray drop size by number (top graph) or by volume (lower graph) is very broad with a number median (nmd or $D_{N.5}$) dia. of 215 μm and volume median (vmd or $D_{v.5}$) of 1190 μm. Figure 4 shows a similar jet nozzle (0.32 cm orifice) operated across the 161 km/hr airstream. Now the size range is actually reduced, but of course, so also is the vmd to around 340 μm and nmd to less than 56 μm [3].

The tube-jet nozzle has been proposed as a device to produce a narrower drop size range. However, Figure 5 for a 0.28 cm inside dia. tube operated at 276 kPa at 0° or with a 161 km/hr airstream shows a vmd of 595 μm and nmd of 69 μm or somewhat smaller than the plate orifice of Figure 3, but still a very broad drop size range.

A hollow cone (disc-core) type nozzle drop size characteristics are shown in Fig. 6. Here the 0.24 cm orifice has a #46 core or whirl plate which produces the tangential spin to give a hollow cone distribution. The nmd is 263 μm while the vmd is 435 μm, with a similar range of the jet across or 90° to the airstream of Figure 4. Note on these drop size figures that a set of curves or drop size vs. cumulative volume and number is shown in the upper right of the figure.

Narrowed Drop Size Range

It is generally agreed among those working with pesticides that the broad range of drop size shown for the nozzles above produce an undesirable loss of active chemical due

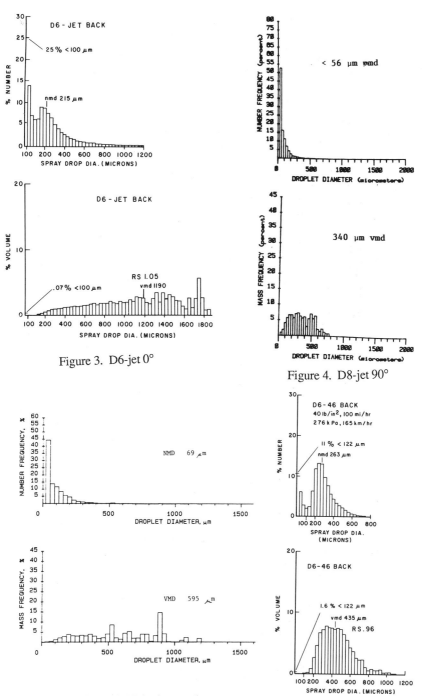

Figure 3. D6-jet 0°

Figure 4. D8-jet 90°

Figure 5. Tube jet nozzle

Figure 6. D6-46 nozzle

to sizes both too large for efficient plant coverage and too small for contact with plants as well as being prone to airborne movement or drift-loss [10]. Thus, there have been various investigations designed to reduce the drop size range and to approach a monodisperse or one drop size only atomization [6]. While various laboratory devices have been developed for this purpose quite successfully, the only commercially available device is that shown in Figure 7 for the Microfoil nozzle. This consists of a row of tubes of 0.033 cm internal dia. arranged on the trailing edge of an airfoil. When operated at less than 35 kPa pressure and up to around 100 km/hr, the output of drops will appear as shown in Figure 8 where a vmd of 643 μm and a nmd of > 56 μm is shown. As can be seen, the size range is a considerable improvement over the larger orifice or tube type nozzles. But limitations on usefulness are subject to using clean liquids which will pass through the small needle-like orifices.

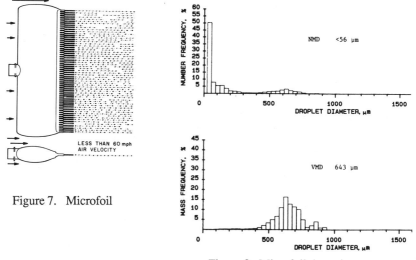

Figure 7.   Microfoil

Figure 8. Microfoil drop size

Rotary Atomizers: Various types of rotary disc, screen and porous cylinders have been used to spin off drops from the periphery of such surfaces [11]. When operated at very precise flow and disc speed, a very narrow drop size range may be achieved. But under normal conditions of aircraft use, the characteristics of a device such as shown in Figure 9 will produce drops of a range of sizes as appear in Figure 10. Here the rotary velocity was 1500 rpm or around 500 m/min peripheral velocity. The vmd was 410 μm and the nmd 81 μm or in the range of the pressure atomizer D6-46 (Figure 6).

Formulation Effects on Drop Size

The basic physical factors of surface tension, viscosity and density, can be related to the drop size produced by a given atomizer under specific operating conditions [12]. Generally using any adjuvant additive or a specific pesticide chemical formulation will reduce surface tension by about 50%. Thus, water at 72 mN/m becomes 35 to 40 mN/m when the active pesticide is mixed into a water carrier. This surface tension reduction decreases the drop size from a given atomizer but is also dependent on the type atomizer as will be shown. Viscosity has less effect on drop size, and appears to alter the flow rate through the atomizer rather than acting physically on the liquid being broken up.

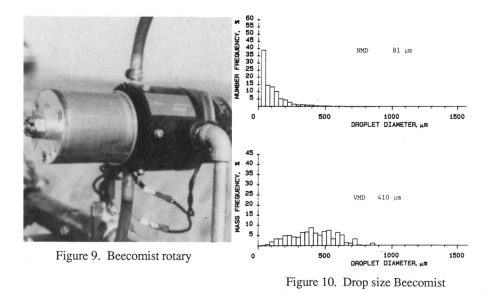

Figure 9. Beecomist rotary

Figure 10. Drop size Beecomist

Visco-elastic additives such as water soluble polymers have a complex reaction appearing to bind the water carrier to some extent, thus increasing drop size [13]. However, again it is dependent on the atomizer, the polymer used and, of course, on the quantity of the polymer agent. Wettable and flowable powders increase viscosity to a small degree in the amounts used in pesticide formulations. The density of the product may change, but when mixed with the carrier water, it produces only a small density difference compared with water. Thus, the most significant changes in drop size come from surface tension reduction with viscosity effective only in large changes of this factor.

The data collected over the past 10 years by the agricultural engineers at Davis has been graphed as shown in the following figures for fan, cone, jet and rotary atomizers. Figs. 11 through 16 are plotted through data points, not with regression analysis. Figure 11 shows data for fan type nozzles plotted for drop size as vmd vs. flow rate in l/min. Two fluids are shown, water and a cottonseed oil-stoddard solvent (85 and 15% by vol.) mixture. The surface tension of the latter is 37 mN/m while viscosity is 19 mP.S and density 0.92 gm/ml. The effect of discharge angle to the airstream (161 km/hr) is also shown, while liquid pressure was maintained at 276 kPa. The largest drops are produced with water at 0° to the airstream followed by CSO (cottonseed oil mix) at 45°, then water at 90°, CSO at 90°, CSO at 0° and smaller drops with water, and lower still CSO at 135° to the airstream. Changing from 0° to 90° with water caused a large change while little change occurred with CSO at 90° and 0°. The large drops with CSO at 45° simply indicates the complex nature of the air action on the fan discharge; the CSO narrowed the fan angle considerably.

Figure 12 shows drop size vs. flow rate for hollow cone disc-core type nozzles, again with water and CSO mixture. Now the CSO produced larger drops than did water at 0° and practically the same size drops at 90° to the airstream. Since the hollow cone produces a circular pattern the effect on nozzle angle is less complex than for the fan. Figure 13 shows the jet nozzles and also recirculating types tested (as Delavan Raindrop® type). Here an adjuvant was used at around 0.03% by volume resulting in 35 mN/m surface tension and 1.2 mPa.S viscosity and 0.98 gm/ml density. The results are straight forward, for decrease in size of drops with adjuvant at 0° to the air and decrease of drop size at 90° vs. 0° with water alone.

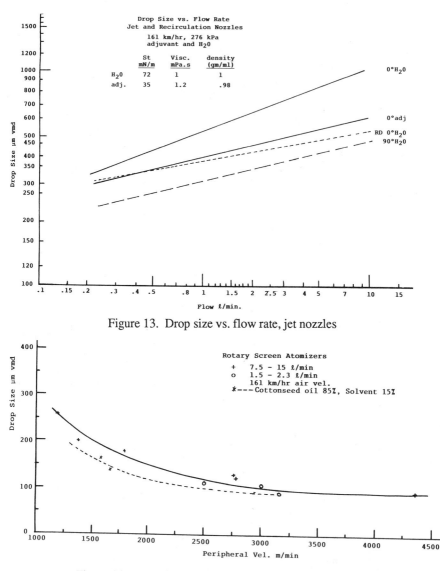

Figure 13. Drop size vs. flow rate, jet nozzles

Figure 14. Drop size vs. peripheral velocity, rotary nozzles

flooded, then the drops are no longer spun off in individual streams (Raleigh break up) but become "sheet like" in form resulting in a larger drop size. Figure 15 shows the almost minimal effect of viscosity change on drop size with two formulations by bacillus Thuringiensis. The Thuricide® is in a water base and the Dipel® is in an oil carrier. The viscosity scales are different for each, but variation of concentration of the bacillus while increasing the viscosity, had little effect on drop size produced by the fan nozzle 8004 operated at 90° and 135° to the airstream of 161 km/hr with 276 kPa liquid pressure for each. Brookfield viscometer shear rate was 50-300 reciprocal seconds.

Figure 16 shows the effect of adding a polymer material to 378 l of water plotted against the drop size produced. Two nozzles are shown, an 8006 fan and a D8-46

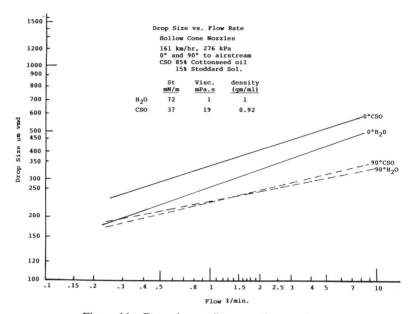

Figure 11. Drop size vs. flow rate, fan nozzles

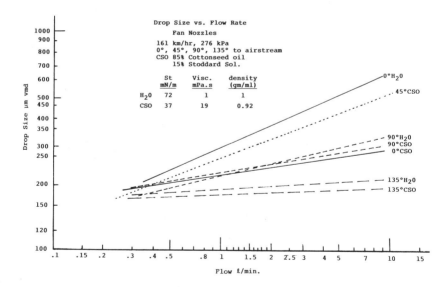

Figure 12. Drop size vs. flow rate, hollowcone nozzles

Rotary screen atomizers are also affected by formulation as well as by peripheral velocity of the rotating screen. Figure 14 shows drop size vs. peripheral velocity ml/min for 2 formulations, water and cottonseed oil, as noted previously. Here the oil resulted in smaller drops, as does also for an increase in peripheral velocity. It is to be noted that limits of flow rate would also have to be imposed here, when the screen becomes

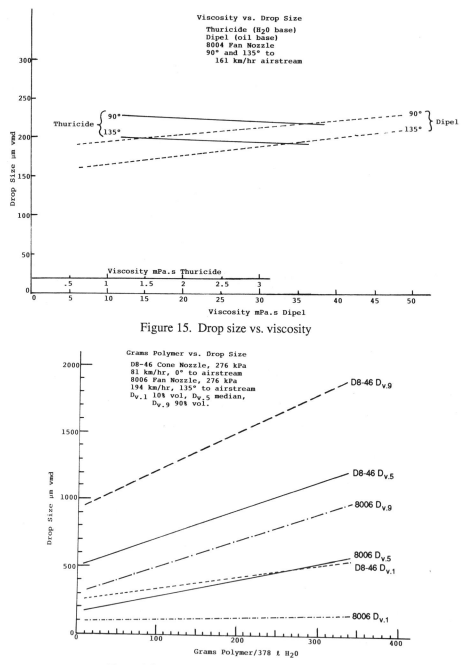

Figure 15. Drop size vs. viscosity

Figure 16. Drop size vs. polymer concentration

hollow cone. Also shown are the vmd (50% size or $D_{v.5}$) as well as $D_{v.9}$, large end (90%) of drop spectrum and $D_{v.1}$, or small spectrum (10%) end. Keep in mind that this was with water only, no pesticide or adjuvant other than the polymer was used. The slope of the curves is indicative of the effect the increasing polymer had in drop size. First, the solid lines show $D_{v.5}$ or median for each nozzle, with positive effect, somewhat alike, for increased median size as polymer increased. Secondly, the increase in $D_{v.9}$ end of drop spectrum also shows a positive or strong reaction to the polymer. But the $D_{v.1}$ or small drop sizes show least effect or reaction to polymer addition. This is significant in that large drops appear to be increased in size, or numbers reduced while smaller drops are less affected, i.e., not increased in size. But again, this is more of an ideal situation, the addition of a pesticide or other adjuvants to the mixture can result in more small drops being produced [1, 13].

MODELLING PESTICIDE APPLICATION

Drop Size Modelling

Utilizing the drop size data collected over several years from the PMS and wind tunnel facility, we have plotted drop size measured (as indicated on the bottom scale of Figures 17, 18 and 19) vs. the regression line or predicted drop size. All data is from water or water base formulations. Figure 17 is shown for fan type nozzles and the predictive equation is vmd = 4073 $(fr^{0.17})(sa^{-0.03})(v^{-0.58})$ where fr is flow rate in gal/min, sa is angle in degrees of discharge to the airstream and v is the airstream velocity (mi/hr). As can be seen, there is considerable scatter of the measured points, but this appears to be a reasonable first cut evaluation.

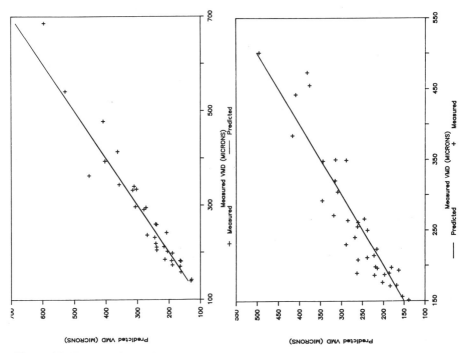

Figure 17. Fan nozzle, vmd predicted vs. measured

Figure 18. Hollow cone nozzle, vmd predicted vs. measured

Figure 18 is shown for the hollow cone nozzles where the regression equation for the predictive curve is: vmd = 4168$(fr^{0.22})(sa^{-0.014})(v^{-0.402})$. Again, the measured drop sizes show considerable scatter and it is expected, as we reexamine some of these, changes in our equation could result.

Figure 19 is for the jet and recirculation nozzles and the equation is: vmd = $107,151(fr^{0.205})(sa^{-0.027})(v^{-1.24})$. Here the data points appear to provide reasonably tight relation to the predictive equation.

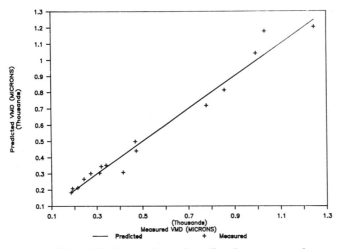

Figure 19. Jet nozzle vmd predicted vs. measured.

<u>Deposit and Drift-Loss</u>

Carrying the drop size data forward and combining it with (a) weather information, and (b) aircraft characteristics, results in the graphs of Figure 20 and 21, where Figure 20 shows the downwind measured drift level points taken from actual field data runs. The line or curve is for a predicted downwind fallout recovery based on the following:

<u>Aircraft Operational Characteristics</u>: (a) aircraft velocity (ground speed) range, 50, 75, 100 and 125 mi/hr, (b) spray release height, range 0 to 40 ft, (c) flagged swath width 1 to 1000 ft, (d) application total vol. 1 to 100 gal/ac.

<u>Weather Parameters:</u> (a) stability ratio $(t_{10} - t_1/U_5{}^2)$ 10^5 where t_{10} and t_1 are temperatures (C°) at 10 and 1 m heights and U^2 is wind velocity in cm/sec at 5 m height. Value range -2.5 to 2.5, wind vel. 0 to 16 km/hr.

The above information, along with the drop size data, are used to develop the graphs shown in Figure 20 which indicates downwind fallout from the given application and for Figure 21 which indicates the airborne concentration resulting from the given application. This data is a "first try" at fallout and drift-loss prediction and should be used as a guide and not for precise results from a given application.

SUMMARY

We have tried to develop a field useful type program, both as a computer data base and as a hard copy manual which will provide some basic information for pesticide applicators, regulatory agencies and manufacturers and dealers of pesticide materials. The EPA registration of new pesticides as well as reregistration of older ones requires a basic study of the potential contamination and resulting damage to wildlife and humans

Log(Measured Fallout)

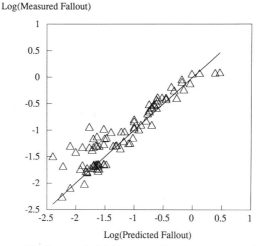

Log(Predicted Fallout)

Figure 20. Downwind fallout, predicted vs. measured

Log(Measured Drift)

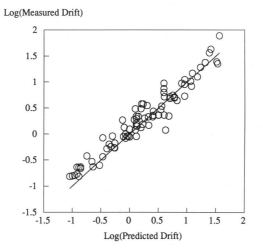

Log(Predicted Drift)

Figure 21. Downwind airborne, predicted vs. measured

and animals, as well as nontargeted crops in the area of the spray application [14]. The data on application requests from the registrant first, data on drop size proposed for use of the material and, secondly, a full-scale drift-loss test if the material is deemed to require this level of examination. This information should be useful in guiding the registrant as well as the agency in a closer evaluation of potential problems from pesticide chemicals.

The program PARIS will permit, first, a calibration mode which aids the user in determining what atomizers are available for the drop size and flow rate desired. The second or drift-loss mode guides the user to define the application constants and to estimate the downwind drift-loss (airborne) and fallout that would result. As a corollary to this mode a maximum value of fallout or airborne concentration can be inputed and the application constants found that will permit the application to be made without exceeding these values. Obviously this program is far from perfection, but as it is further developed and used, we hope it can be refined to make it more useful and accurate and enable continued safe and effective pesticide use.

REFERENCES

[1] Akesson, N. B., and Yates, W. E., "Pesticide Deposit and Activity as a Function of Spray Atomizers and Liquid Formulations," Proceedings 11th International Congress on Agricultural Engineering, CIGR Secretariat 22 Clyde Rd., Dublin 4, Ireland, 1989.

[2] Akesson, N. B., and Yates, W. E., "Physical Parameters Affecting Spray Application," Chemical and Biological Controls in Forestry, Willa Y. Garner and John Harvey, Jr., Eds. American Chemical Society Symposium Series #238, Washington, D.C., 1984, pp. 95-115.

[3] Yates, W. E., Akesson, N. B. and Cowden, R. E., "Measurement of Drop Size Frequency from Nozzles Used for Aerial Applications of Pesticides in Forests," USDA, Forest Service Equipment Development Center, Missoula, Montana, 3400 F PM, 1984, 8434-2804.

[4] Hall, F. R., "Effect of Formulation, Droplet Size and Spatial Distribution on Dose Transfer of Pesticide," Pesticide Formulations and Application Systems: 10th Volume, ASTM STP 980, D. A. Hovde and G. B. Beestman, Eds., American Society for Testing and Materials, Philadelphia, 1989, pp. 145-154.

[5] Derksen, R. C. and Bode, L. E., "Performance Characteristics of Four Rotary Atomizers," Pesticide Formulations and Application Systems: 6th Volume ASTM STP 943, D.I.B. Vander Hooven and L. D. Spicer, Eds., American Society for Testing and Materials, Philadelphia, 1987, pp. 68-87.

[6] Yates, W. E., Cowden, R. E. and Akesson, N. B., "Nozzle Orientation, Air Speed and Spray Formulation Effects on Drop Size Spectrums," Transactions of American Society of Agricultural Engineers, St. Joseph, Michigan, Vol. 26(6), 1983, pp. 1638-1643.

[7] Akesson, N. B. and Gibbs, R. E., "Precision and Safety with PARIS: Some Basic Challenges for Aerial Application," American Society of Agricultural Engineers, St. Joseph, Michigan, Paper Number AA-ASAE 88-008, 1988.

[8] Sandaram, A., Ratnakaren, A., Raske, A. G. and West, R. J., "Effect of Application Rate on Droplet Size Spectra and Deposit Characteristics of Dimilin[R] Spray Mixtures in an Aerial Spray Trial," Pesticide Formulations and Application Systems, 7th Volume, ASTM STP 968, G. B. Beestman and D.I.B. Vander Hooven, Eds., American Society for Testing and Materials, Philadelphia, 1987, pp. 104-115.

[9] Knollenberg, R. G., "The Optical Array: An Alternative to Scattering or Extinction for Airborne Particle Size Distribution," Journal of Applied Meteorology, Vol. 9(1), 1970, pp. 86-103.

[10] Akesson, N. B., and Yates, W. E., "Development of Drop Size-Frequency Analysis of Sprays Used for Pesticide Applications," Pesticide Formulations and Applications Systems 5th Volume ASTM STP 915, Thomas M. Kaneko and Larry D. Spicer, Eds., American Society for Testing and Materials, Philadelphia, 1986, pp. 83-93.

[11] Bode, L. E. and Zain, S. B., M.D., "Spray Deposits from Low Volume Applications Using Oil and Water Carriers," Pesticide Formulations and Application Systems, 7th Volume ASTM STP 968, G. B. Beestman and D.I.B. Vander Hooven, Eds., American Society for Testing and Materials, Philadelphia, 1987, pp. 93-103.

[12] Haq, K., Akesson, N. B. and Yates, W. E., "Analysis of Droplet Spectra and Spray Recovery as a Function of Atomizer Type and Fluid Physical Properties," Pesticide Formulations and Applications Systems 3rd Volume ASTM STP 828, American Society for Testing and Materials, Philadelphia, 1983, pp. 67-82.

[13] Bouse, L. F., Carleton, J. B. and Janks, P. C., "Effect of Water Soluble Polymers on Spray Droplet Sizes," <u>Transactions of American Society of Agricultural Engineers,</u> St. Joseph, Michigan, Vol. 31(6), 1988, pp. 1633-1641, 1648.

[14] Holst, R. W., "Pesticide Spray Evaluation: Droplet Size Spectrum, Test and Drift Field Evaluation Tests," <u>EPA Office of Pesticides Programs,</u> Hazard Evaluation Division, EPA 540/9-86-131, 1986.

Franklin R. Hall, Joseph P. Reed, Donald L. Reichard, Richard M. Riedel and Jeffrey Lehtinen

PESTICIDE DELIVERY SYSTEMS: SPRAY DISTRIBUTION AND PARTITIONING IN PLANT CANOPIES.

REFERENCE: Hall, F. R., Reed, J. P., Reichard, D. L., Reidel, R. M. and Lehtinen, J., "Pesticide delivery systems: Spray distribution and partitioning in plant canopies." Pesticide Formulations and Application Systems: 10th Volume, ASTM STP 1078, L. E. Bode, J. L. Hazen and D. G. Chasin, Eds., American Society for Testing and Materials, Philadelphia, 1990.

ABSTRACT: Considering only a small fraction of pesticidal sprays reaches the target, more attention needs to be placed on developing techniques which increase crop canopy penetration. Quantification of where pesticides are going is clearly going to be emphasized by EPA as a mandatory process for all future registrants. Drift represents a loss of chemical and implies dangers of air and water pollution. A critical problem is orchard airblast sprayer applications. Plant coverage by 3 boom sprayers was evaluated on tomatoes. Partitioning and overall canopy distribution, was measured by % area covered via image analysis. Similiarly, partitioning and distribution of sprays from a modified airblast sprayer was evaluated using water sensitive paper. On field grasses, paraquat applied at recommended field rates was used to measure efficacy and drift damage with the electrostatic nozzle charged vs. a conventional hydraulic nozzle. In short grass, charged was superior to uncharged whereas in long grass, both were superior to other treatments. In the laboratory, drift was reduced by 99.0% by shrouding and charging electrostatic (ENS) nozzles. The ramifications of sampling method, distribution inconsistencies, and biological effects are discussed. These results in combination with capture efficiency studies will allow future studies on spray accountancy to proceed in accordance with EPA SOP standards (40CRF, 158.142).

KEYWORDS: Atomizers, canopy partitioning, image analysis

Professor Hall is Head of the Laboratory for Pest Control Application Technology (LPCAT), Mr. Reed is a graduate research associate, Mr. Reichard is an agricultural engineer USDA/ARS, Professor Reidel is a plant pathologist, The Ohio Agricultural Research and Development Center, The Ohio State University, Mr. Lehtinen is an engineer with Parker-Hannifin Corp. Cleveland, OH.

Trends and developments in pesticide application technology are dynamic. Currently, low volume spray applications are replacing high volume spraying of field and tree crops [1]. Hence, the understanding of factors which govern the trajectory, velocity, and eventual size of droplets produced by high volume spray atomizers may be obsolete [2]. Correspondingly, a change in the characteristics of the crop canopy may have a profound influence upon the efficiency of spray application [3]. In essence, the confounding of these two factors, spray application volume and crop canopy, presents a new set of conditions that may not reflect current trends.

Earlier studies indicated that high volume applications could not be relied upon to provide proper coverage [4]. Although this may have been the impetus to research and subsequently use lower application volumes, each application method has its own set of protocols for maximizing deposition. Further, comparisons between the spray application systems are difficult to assess since vitally needed details in the literature are sometimes ommitted [5]. Also, protocols are lacking for standardization of spray application parameters. Thus, the previously stated criticisms lend further credence to the question of what deposit criteria are needed at various target sites for the most efficient control of pests.

Spray droplet deposition in a crop canopy is influenced by droplet size, velocity, and formulation [6]. Therefore, the potential to manipulate these factors to optimize the efficiency of drop formation and impaction upon the plant is great [7]. Perhaps the easiest factor to manipulate is the addition of a surfactant to increase spray retention [8]. Anderson and Hall [9] reported the addition of surfactant was related to dynamic surface tension and retention. Once the quantity of added surfactant exceeds both the critical micelle concentration and dynamic surface tension of 26 dynes/cm, retention markedly improves.

The retention of spray droplets by manipulation of droplet size is not clearly understood. For example, Spillman [10] suggested spray droplets <100 um diameter would be retained regardless of any other variables. However, high speed cinematography has demonstrated water droplets with 67 μm diameter rebounded as easily as larger sized droplets [11]. Thus, the dynamic surface tension of the spray solution is important when considering the retention of spray droplets.

Velocity of the droplet also plays an important role. When droplet velocity possesses a kinetic energy which exceeds a threshold that is equal to/or greater than that of droplet impaction and spread, rebound occurs. This may be beneficial, especially in crops where the upper leaf surface is waxy and does not retain droplets but the lower surface is not waxy and retains droplets [12]. However, droplets not retained on the plant may contribute to either spray drift or impingement on the soil or in water, both of which are environmentally undesirable [13].

Distribution of the 'spray cloud' within the crop canopy is complex. Of particular interest is the vertical distribution of spray droplets in the crop canopy. Although Gohlich [14] characterized the DV.5 of a nozzle at ground, middle and top of a wheat canopy, larger droplets were distributed more towards the top of the crop canopy. Smaller droplets were found closer to the ground. Further, Bache [15] modelled the number distribution of not only droplets in a crop canopy but the size of the droplets as well.

Undoubtedly no clear cut answers can be provided for the question, "What is the ideal droplet size?" [16]. However, the ability to manipulate droplet size via spray application machinery provides 'tools' that can be used to elucidate further upon the ideal droplet size. In general, the hydraulic nozzles used for pesticide spraying produce droplets with a wide range of droplet sizes [17]. Reduction of unwanted droplet sizes led to the development of rotary and controlled droplet applicators (DA) which produce a more narrow droplet spectrum [18]. Further, the discovery of accurately placing smaller droplets by the electrostatic charging of individual droplets has been reviewed by Hislop [19]. The drawbacks of the alternatives to the conventional hydraulic nozzles has been a question of economics and whether these new methodologies do what their manufacturers claim (i.e., better pest control, reduce drift, etc.).

Realization of the biological significance of spray droplet retention and distribution throughout the crop canopy is just beginning to be understood. First, whether the target pest is an insect, plant pathogen or weed will determine where the pesticide must be applied in the crop canopy. Further, this criteria can be modified by the mode of action and sytemic activity of the pesticide for the crop. For example, spray coverage of weeds by the systemic postemergent herbicides, fluazifop-butyl, haloxyfop, and sethoxydim did not improve efficacy as well as the addition of various adjuvants [20]. In contrast, paraquat, a non-systemic herbicide displayed greater penetration and phytotoxicity within a weed canopy when applied by fan and cone atomizers than CDA's [21]. In the case of insects and plant pathogens, spray coverage appears to be an important factor in pesticide efficacy. Adams and Palmer [22] found air assistance was required for placing charged droplets within the crop canopy. Further, they found the combination of air assistance and small droplets within a narrow droplet size spectrum was required for a uniform coverage of upper and lower leaf surfaces of greenhouse tomatoes. In a similiar fashion, percent area covered was an important criterion in the efficacy of dinocap to control powdery mildew in apples [23].

In general, studies dealing with crop/weed canopy penetration by pesticides as evidenced in the prior examples allude to the deposit parameter, percent area covered. The objective of this research was to compare partitioning and canopy coverage in tomatoes, apples and grasses by various application machinery.

MATERIALS AND METHODS

Tomato Canopy Study

In 1988 and 1989, field studies were conducted at Fremont, OH on the vegetable crops research station. The two major objectives of this exercise were to determine whether spray application equipment had a major effect upon the seasonal protection of tomatoes. If differences existed between application machinery, then the second objective was to ascertain whether the spray coverage by the application machinery was responsible for the differences. Georgia grown tomato transplants H1810 were planted on May 18 and May 24 for 1988 and 1989, respectively. Each plot consisted of one row that was 13 m long and plants were spaced 30 cm apart. Border rows were on both sides of the plots. The blocks were separated by a 4 m alleyway and there were four replicates.

All plots were sprayed on a two week time schedule. The first sprayer was a tractor mounted (International Harvester 140) small plot sprayer calibrated to deliver 618 l/ha of spray solution at 3.2 km/h. Four hollow cone nozzles (Delavan HC-8) straddled one row, nozzle spacing was 18.3 cm at a height of approximately 45 cm above the canopy. A CO_2 pressure system was used to apply approximately 414 kPa at the nozzles.

The second sprayer was a high volume field sprayer (FMC DO 35P/500S) that was also calibrated to deliver 618 l/ha at 550 kPa. FMC nickel alloy, hard core nozzles with #2.5 discs were used for the study. The travel speed was 6.4 km/h and nozzle spacing was 30 cm; only the nozzles of the middle boom section were operated at a height of ca.75 cm.

The third sprayer was an air assist Hardi model (Mini-Variant). At the end of each 10 cm diameter flexible hose is a single nozzle around which air was forced from a centrifugal fan to disperse spray. Both the fan and pump are power-take-off operated. Two nozzle/air spout were centered over each row at a height of ca. 40 cm. The operating pressure was 690 kPa and the sprayer was calibrated to deliver 330 l/ha of finished spray at 6.4 km/h.

Application of fungicides, chlorothalonil (Bravo) early in the season followed by late season application of metalaxyl and chlorothalonil (Ridomil-Bravo), were applied at the following rates: 600 g of chlorothalonil per 95 l of water; and 454 g of chlorothalonil-metalaxyl in 95 l of water. Mixing and agitation was accomplished by bypass agitation in both the FMC and Hardi sprayers. The small plot sprayer relied upon the method of hand mixing and once loaded upon the tractor, general machine movement.

Water sensitive spray cards (Spraying Sysytems Co., Wheaton, IL) were placed on both the top and bottom of tomato leaves at various heights from the ground (top 60 cm, middle 30 cm and ground 0 cm) along the stem area and near the periphery of the plant. This enabled an assessment of tomato canopy penetration by the various sprayers (Figure 1). At each position, two cards (2.5 x 7.6 cm) were stapled together to sample deposition on both sides of a leaf. Water sensitive spray cards were set out on July 22 and July 24 for analysis of percent area covered for 1988 and 1989, respectively. A Dapple Systems II image analyzer was used to count droplets captured on the spray cards and calculate the percent area covered as well as the largest and smallest droplets. Three randomly located measurements on each spray card were taken to obtain the final replicate mean per sample site. All percent area covered data was subjected to an analysis of variance procedure following an arcsine transformation. Mean separation procedures involved the use of Duncan's multiple range procedures at the $p = 0.05$ level.

Apple Canopy Study

Two apple tree canopies representing both standard and dwarf canopy cultivars (Red and Golden Delicious), were used to assess differences in canopy penetration for four different airblast sprayer configurations during 1987. Canopy dimensions for the Red Delicious was 10.2 m high by 9 m wide while the Golden Delicious was 8.2 m high by 9 m wide. A Maxi-mist FMC airblast sprayer was used in this study. The airblast sprayer fan speeds were 1400 and 2800 rpm for low and high settings. The sprayer also had an additional modification which was a 45° metal shroud (45 cm. long) that sloped backwards. The sprayer was used at both fan speeds as well as with and without a shroud with 6, disc/core, hollow cone spray nozzles with 2 hole whirl plates and was calibrated to deliver 475 l/ha at 930 kPa. The orchard driving alley was 6.6 m wide and for the objective of this study, the sprayer made only one pass parallel to the side of the tree line which was 3.3 m from the centerline.

To determine the canopy penetration of sprays, study poles were placed at three positions. The first pole was located near the trunk, the second was within the tree row, but on the periphery of the tree. The third pole position was on the other side of the sprayer path perpendicular to the centerline (Figure 2). Water sensitive spray cards (2.5 x 7.6 cm) were placed up to 5 m at 1, 2, 3 and 4 m increments on each pole. Two cards were attached at each height with fasteners, to represent front (facing sprayer) and back side of leaves.

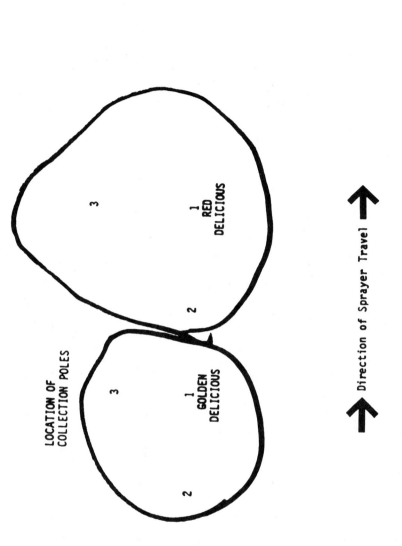

FIGURE 2. Location of water sensitive spray card collection poles throughout Red and Golden Delicious Tree Canopies.

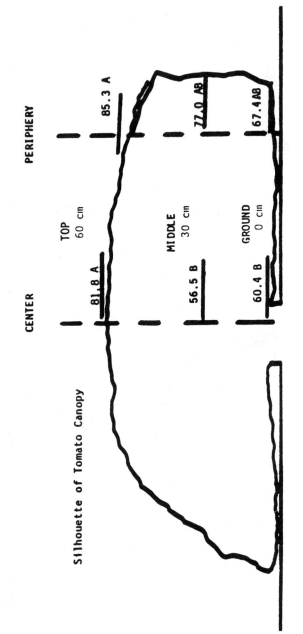

FIGURE 1. Location of water sensitive spray cards in the tomato canopy and percent area covered by a hydraulic sprayer application at various locations. Means following the same letter are not significantly different at the 0.05 level of significance. LSD = 19.6.

The experiment was analyzed as a split-split plot design which was replicated three times. Each pass represented a replicate. After each pass, all water sensitive spray cards were collected and new cards installed at each site. The main plots were sprayer, sub-plots were height, sub-sub plots were position and sub-sub-sub plots were side of leaf. All data was subjected to analysis of variance procedures. Percent area covered was transformed to arcsine values before analysis of variance. Means were separated by Duncan's multiple range procedure at the p = 0.05 level.

Field Grass Study with ENS Atomizer

A comparison between the ENS electrostatic nozzles of various charging and shroud configurations was established to determine the efficacy of the herbicide, paraquat, in tall and short grass as an assessment of the application parameter. The primary grass weed was fall panicum, Panicum dichotomiflorum and heights were 60 and 15 cm for tall and short canopies, respectively. Since paraquat is non-systemic, this enabled us to utilize a visual monitoring evaluation scheme for off-target drift as another parameter for estimating the application efficiency of the various atomizers. An opposing nozzle configuration was found to be the most efficient method to apply herbicide solutions through the ENS nozzle [24]. The ENS nozzle was calibrated to deliver 86 l/ha at 3.2 km/h. The output per nozzle was 72 l/ha, spacing was 15 cm. Four nozzles were mounted in a plexiglass shield; two of which opposed one another at 45° angles from the horizontal tool bar. The skirt consisted of a rubberized cloth material with slits to accommodate movement over uneven surfaces and was bolted around the entire shield and extended 15 cm down to the ground. Liquid and air pressures were 103 kPa, and the nozzle inductively charged the droplets with 1000 v. The charge mass ratio was 1.5 mC/Kg using a Faraday cage device [24]. An 80° flat fan hydraulic nozzle was used for comparison in this study as a standard. The desired application rate was 190 l/ha at 4.8 km/h travel speed with 50 cm nozzle spacing. The flow rate was 75.7 l/min at an operating pressure of 276 kPa. Paraquat was applied at the recommended field rate of 0.28 kg/ha.

The experimental design was a randomized complete block design that was replicated three times. The treatments were as follows: charged spraying with and without skirt; uncharged spraying with and without skirt; and the hydraulic nozzle, for a total of 5 treatments. Fourteen days after treatment, plots were visually rated for percent weed control. All data was transformed before analysis of variance and mean separation was performed using Duncan's multiple range test at the p = 0.05 level.

ENS Laboratory Drift Study

    In the laboratory, a study was initiated to establish the effect of charging and use of a skirt to reduce spray drift. The various configurations of the sprayer shield were the main plot of a split plot design. Wind speed was the sub-plot. The experiment was replicated four times. Water sensitive spray cards (2.5 x 7.6 cm) were placed at 60 cm increments up to 3.6 m down wind from the sprayer apparatus with fans operated to deliver air at 0, 3.2, and 6.4 km/h. Wind was provided by the use of two rheostat controlled 500 watt fans. The fans were placed ca. 180 cm from the shield. Hydraulic and ENS nozzle sprayer operating parameters were as noted in the field study. However, the sprayers were operated only for 90 seconds so the spray cards would not become overloaded with water deposits. All water sensitive spray cards were analyzed for percent area covered. Data were transformed to arcsine values prior to analysis of variance procedures. Mean separation (using Duncan's multiple range test) was used to delineate any significant differences at the $p = 0.05$ level.

RESULTS AND DISCUSSION

Tomato Canopy Study

    Canopy partitioning comparisons of the spray coverage by various sprayers is presented in Table 1. The small plot sprayer and FMC sprayers demonstrated both lower coverage values and more variability throughout the canopy. All three sprayers exhibited poor penetration of the ground center portions of the canopy. The Hardi sprayer had significantly better coverage on the ground periphery areas, compared to both the small plot and FMC sprayers. The FMC sprayer with the 30 cm spacing delivered significantly greater amounts of spray to the between-row ground areas. In contrast, the Hardi, with targeted (directed) delivery, appeared to place more material into the foliage areas. However, when rows varied in growth/spreading patterns, the spouts, if placed too close to (or too far from) the canopy, could result in sprays missing some of the foliage. This may have contributed to reduced production of red tomatoes in 1988 (Table 2). In 1989, the penetration differences were less variable, possibly because of the significant differences in growth habit - a drought in 1988 and excessive early season rainfall in 1989. Slow drying conditions associated with the many rainy days in 1989 may have contributed to additional run-off. This is an excellent example of meteorological and physico-chemical interactions of spraying, which concurs with the findings of Johnstone [25].

    The influence of sprayer coverage for controlling the tomato disease, blossom end rot (Anthracnose spp.), is shown in Table 2. In 1988, tomato production was greater with the small plot sprayer and FMC sprayer. This trend in production was the opposite in percent coverage values of spray cards. However,

TABLE 1 -- Comparison of droplet deposition on side of leaf for various sprayers.

Sprayer	Leaf Surface	Percent Area Covered	
		1988	1989
Small Plot	Upper	106.7 a	61.9 ab
Small Plot	Lower	78.6 b	14.2 d
Hardi	Upper	77.3 b	64.3 a
Hardi	Lower	71.1 b	51.2 b
FMC	Upper	41.7 c	33.0 c
FMC	Lower	64.9 b	54.8 ab

Means in the column with different letters are significantly different at the p = 0.05 level using Duncan's multiple range test.

production in plots sprayed with the Hardi sprayer was not significantly different than those plots sprayed by the FMC, but production was less than the small plot sprayer. The difference among sprayers may be attributed to the carrier volumes. Recall that the small plot and FMC spray volumes were 618 l/ha whereas the Hardi had an application volume of 330 l/ha, yet the card coverage did not reflect the application volumes. First, the production of fine droplets by the sprayers may have been sufficiently great enough to alter deposition by high spray volumes. This is further supported by the fact that the high operating pressures of sprayers can increase the proportion of fine droplets in the spray cloud. Lastly, 1988 was an extremely dry year and the extra application volume of the FMC sprayer may have contributed to "run-off" and perhaps better coverage.

In 1989, tomato production in FMC sprayer plots was neither significantly different than the check nor the other sprayers (Table 2). Production by both the Hardi and small plot sprayer were significantly different than the check, indicating that in both years disease control was responsible for increased tomato production. In general, coverage appeared to be responsible for increased efficacy.

All sprayers were observed to produce significant off-row movement of spray material (addressed in a separate paper) which in small plot trials may contribute to a confounding of plot effects. The droplet data, as measured by percent area covered, of course is only a reflection of the initial delivery efficacy and does not address the potential for redistribution, which in the case of disease protection by fungicides is an essential component of the plant protection process.

TABLE 2 -- Total harvested red tomatoes for various sprayers applying chlorothalonil/metalaxyl for Anthracnose control.

Sprayer	Production (kg/13 m)	
	1988	1989
Small Plot	64.5 a	51.8 b
FMC	50.0 ab	50.5 b
Hardi	42.1 b	38.2 ab
Check	31.0 c	28.2 a

Means in the same column with a different letter are significantly different at the $p = 0.05$ level using Duncan's multiple range test.

Apple Canopy Study

Vertical coverage comparisons in the two apple canopies are presented in Table 3. Regardless of height, the modified airblast sprayer at the low fan speed provided significantly better coverage than operation at the high fan speed in the Red Delicious (Standard canopy). In the Golden Delicious (Dwarf canopy), a fan speed by height deposition interaction occurred. The modified airblast sprayer operating at a high fan speed, resulted in coverage that was significantly greater at 3 and 4 m than at 1 and 2 meters. However, at a low fan speed, the modified sprayer appeared to produce better coverage at the 1 and 2 meter heights.

The coverage obtained by the operation of the standard airblast sprayer in the Red Delicious plot was significantly greater at all heights except 3 meters at a high fan speed than the low fan speed. Again, in the Golden Delicious plot, the high fan speed with a standard sprayer provided significantly greater vertical coverage than a low fan speed, regardless of height.

Lateral canopy coverage or penetration as indicated by deposits at poles 1, 2, and 3 by the standard and modified sprayers in the two different canopied cultivars is shown in Table 4. The modified sprayer operating at a high fan speed had significantly less lateral coverage than the low speed operation in the Red Delicious tree. Few significant differences in the Golden Delicious tree were observed for lateral coverage when application was made by the modified sprayer. One exception was at position 3 (directly opposite the sprayer path); the higher fan speed did improve coverage by the modified sprayer in the Golden Delicious tree.

When the standard sprayer configuration was used, many significant differences in lateral coverage throughout the dwarf and full canopied trees were observed between the high and low fan speeds. Some notable exceptions are directly opposite the sprayer in the Red Delicious trees with standard canopies (position 3); and within the center line (position 2) and adjacent to another tree in the Golden Delicious trees.

TABLE 3 -- Vertical percent coverage in two apple canopies by modified and standard FMC orchard airblast sprayers operated at two fan speeds.

Fan Speed/ Nozzle Configuration	Height(m)	% Coverage	
		(Dwarf) Golden Delicious	(Standard) Red Delicious
High Speed			
Modified	4	19 de	3 b
	3	22 ef	2 a
	2	7 b	3 b
	1	5 ab	2 a
Standard	4	28 h	16 i
	3	23 g	6 c
	2	30 h	20 j
	1	32 h	24 k
Low Speed			
Modified	4	1 a	15 h
	3	2 a	8 d
	2	22 ef	8 d
	1	12 c	29 l
Standard	4	17 d	11 g
	3	26 f	9 e
	2	3 ab	10 f
	1	16 cd	9 e

Means in the same column with different letters are significantly different at the $p = 0.05$ level using Duncans multiple range test.

To summarize, the coverage afforded spatially by the various sprayer configurations in the different apple canopies is variable (Table 5). At the high fan speed, the standard sprayer provided significantly greater coverage in both canopy types than its modified counterpart. Under low fan speed operation, standard sprayer configuration did not provide as thorough coverage as the modified sprayer in the dwarf canopy. However, the exact opposite was observed in the standard tree canopy, with the standard sprayer configuration giving better coverage than the modified sprayer. Reichard et al. [26] found air velocities

TABLE 4 -- Lateral percent spray coverage in two apple canopies by modified and standard FMC orchard airblast sprayers operated at two fan speeds.

Fan Speed/ Nozzle Configuration	Pole Position	% Coverage	
		(Dwarf) Golden Delicious	(Standard) Red Delicious
High Speed			
Modified	1	7.8 e	3.2 i
	2	20.8 b	3.5 i
	3	11.9 de	1.2 j
Standard	1	36.1 a	24.6 a
	2	35.3 a	21.3 b
	3	14.1 cd	4.8 gh
Low Speed			
Modified	1	9.7 e	14.8 d
	2	16.7 bc	13.3 e
	3	1.4 f	18.4 c
Standard	1	13.0 cd	8.9 f
	2	33.0 a	15.7 d
	3	1.1 f	5.3 g

Means in the same column with different letters are significantly different at the $p = 0.05$ level using Duncan's multiple range test.

of orchard airblast sprayers decreased as travel speed increased. Although sprayer travel speed was constant in this study, perhaps a travel and fan speed nozzle interaction is occurring in the different canopies among the different kinds of nozzle configurations. Thus, it appears that improved spray deposition, as a result of air carrier configurations (high/low fan speed and a modified delivery [45° vs 90°] into plant canopies), would be dependent upon (1) target canopy parameters as measured by leaf area index (LAI) [LPCAT unpublished data], (2) an improved definition of where specific sprays are needed, and (3) interaction with speed of travel as suggested by Gohlich [14], Johnstone [25] and Reichard et al. [26].

TABLE 5 -- A summary of percent coverage in two apple canopies by modified and standard FMC orchard airblast sprayers.

Fan Speed/ Nozzle Configuration	% Coverage	
	(Dwarf) Golden Delicious	(Standard) Red Delicious
High Speed		
Standard	16.9 a	28.5 a
Modified	2.6 d	13.6 b
Low Speed		
Standard	9.9 c	15.7 b
Modified	15.5 b	9.3 c

Means in the same column with different letters are significantly different at the p = 0.05 level using Duncan's multiple range test.

Field Grass Study

Tall field grass coverage studies are presented in Table 6. The greatest coverage of any nozzle was the charged ENS nozzle with a skirt, which was significantly different than all other nozzles. The no-skirt ENS nozzle configurations also provided the excellent coverage and were significantly different than all other nozzles as evaluated by % grass control.

In Table 7, short field grass studies indicated the hydraulic nozzle was as efficient in spray delivery as the skirt charged ENS configuration. Both of these nozzle configurations resulted in significantly greater % grass control than all other combinations and the untreated check. The charged unskirted nozzle provided 73.3% grass control, which was less than the hydraulic nozzle. However, no differences were observed between the charged nozzle with no-skirt and no-charge with no-skirt in % weed control. Thus, use of a skirt appeared to add more consistancy to the delivery efficiency of the sprayer in short canopies, but this was not readily apparent in the taller canopies.

TABLE 6 -- Weed Control in Tall Canopy Grasses (60 cm).

Nozzle Configuration	% Grass Control
Skirt Charged	85.0 e
Skirt No-charge	51.6 b
No-Skirt Charged	76.6 d
No-Skirt No-Charge	78.3 d
Hydraulic	66.6 c
Untreated	0.0 a

Means in the same column with different letters are significantly different at the p = 0.05 level using Duncan's multiple range test.

TABLE 7 -- Mean Weed Control in Short Canopy Grasses (15 cm)

Nozzle Configuration	% Grass Control
Skirt Charged	85.0 c
Skirt No-Charge	73.3 b
No-Skirt Charge	73.3 b
No-Skirt No-Charge	71.6 b
Hydraulic	83.3 c
Untreated	0.0 a

Means in the same column with different letters are significantly different at the p = 0.05 level using Duncans multiple range test.

ENS Laboratory Drift Study

The extent of off-target drift for all nozzles in this laboratory study is shown in Table 8. The results show that the percent area covered is inversely proportional to wind velocity. In addition, coverage values down wind for all windspeeds indicates greater coverage closer to the sprayer (Table 9). Hence, an inverse relationship between deposition and distance from sprayer also exists.

TABLE 8 -- Mean coverage for 90 seconds for various wind speeds from nozzles (pooled).

Wind Speed	% Coverage
0 Km/h	15.1 a
3.2	12.9 b
6.4	12.0 c

Means in the same column with different letters are significantly different at the p = 0.05 level using Duncans multiple range test.

TABLE 9 -- Mean coverage for 90 seconds for various distances down wind from nozzles (pooled).

Position (m)	% Coverage
0	68.8 a
0.6	11.9 b
1.2	5.2 c
1.8	3.8 d
2.4	2.0 e
3.0	1.1 f
3.6	0.5 g

Means in the same column with different letters are significantly different at the p = 0.05 level using Duncans multiple range test.

To demonstrate the influence of skirt, canopy and charging with respect to wind speed, coverage values are presented in Table 10. No significant differences were observed between no-shield and no skirt treatments which demonstrated greater off-target coverage in a 6.4 km/h wind at 2 m downwind than either the ENS with a shield and skirt and the hydraulic nozzle. However, ENS nozzle with both a shield and skirt demonstrated significantly better drift reduction (ca. 99.0%) than the hydraulic nozzle. Thus, the use of electrostatics in the field may require the use of devices which reduce off-target drift. Such devices may take the form of a shield or use of air assistance to place droplets where the electrostatic charge may be utilized to greatest benefit.

TABLE 10 -- Mean coverage by various nozzle configurations downwind 2.0 m in a 6.4 Km/h wind.

Nozzle Configuration	% Coverage
No-Skirt	8.49 c
Skirt	0.05 a
No-Shield, No-Skirt	8.88 c
Hydraulic	5.29 b

Means in the same column with different letters are significantly different at the p = 0.05 level using Duncans multiple range test.

This study shows the potential to reduce drift by the use of a shield with skirt and the use of smaller, more biologically active droplet sizes. The use of a skirt around the periphery of the shield allows adequate time and space for the droplet to be captured while still in a liquid state. CAD programs, such as FLUENT (Creare. X Inc.), now being utilized at LPCAT, allow the simulation of velocity, droplet size and air turbulence interactions. The knowledge of these interactions can be used to minimize off-target movement of droplets from electrostatic sprayers which require a smaller range of droplet sizes. The agriculturalist may thus be provided with a wide "weather window" in which to more optimally place crop protection agents. As concluded by Hislop [19], sprayer innovations like electrostatics may thus be utilized to augment conventional systems in our goal of a more accurate pesticidal spray delivery. Concerns (perceptions) about pesticides in the environment should thus aid the search for a more environmentally acceptable technology.

ACKNOWLEDGMENTS

The authors wish to express sincere thanks to Dr. Roger Williams, Dr. Robert Treece and Dr. Harvey Krueger for their review of the manuscript. This manuscript was a contribution of the Laboratory for Pest Control Application Technology, The Ohio State University and the Ohio Agricultural Research and Development Center, Journal Article Series No. 76-90.

[REFERENCES]

[1]   Mowitz, D. Air assist application: Next revolution in spraying. Successful Farming 85, 1987, p 26.

[2]   Young, B. W. Studies on the retention and deposit characteristics of pesticide sprays on foliage. Paper-1-4: Application meeting, IXth CIGR Congress, East Lansing, MI. 1979.

[3]   Hall, F. Canopy volume concepts in pesticide application saves dollars. In Proceedings 131st Meeting Illinois State Horticultural Society, 1987, pp 38-44.

[4]   Akesson, N. B. and W. E. Yates Problems relating to application of agricultural chemicals and resulting drift residues. Annual Review of Entomciogy, 9, 1964, pp 285-318.

[5]   Cayley, G. C., P. E. Etheridge, R. E. Goodchild, D. C. Griffiths, P. J. Hulme, R. J. Lethwaite, B. J. Pye and G. C. Scott. Review of the relationship between chemical deposits achieved with electrostatically charged rotary atomizers and their biological effects. British Crop Protection Council Monograph 2, 1970, pp 23-33.

[6]   Maybank, J. Off-target drift from ground-rig pesticide sprayers, Abstract 113, Weed Science Society of America., Champaign, IL. 1989.

[7]   Reichard, D. L. Drop formation and impaction upon the plant. Weed Technology, 2, 1988, pp 82-87.

[8]   Lake, J. R. and J. A. Marchant The use of dimensional analysis in a study of drop retention on barley. Pesticide Science, 14, 1983, pp 638-644.

[9] Anderson, N. H. and D. J. Hall The role of dynamic
     surface tension in the retention of surfactant sprays
     on pea plants. 1st International Symposium of Adjuvants
     and Agrichemicals, Brandon, Canada. 1986.

[10] Spillman, J. J. Spray impaction, retention and
     adhesion: an introduction to basic characteristics.
     Pesticide Science, 15, 1984, pp 97-106.

[11] Reichard, D. L., R. D. Brazee, M. J. Bukovac, and R. D.
     Fox. A system for photgraphically studying droplet
     impaction on leaf surfaces. Transactions of
     the American Society of Agricultural Engineers, 29,
     1986, pp 707-713.

[12] Ennis, W. B., Jr., R. E. Williamson and K. P.
     Doeschner. Studies on spray retention by leaves of
     different plants. Weeds, 1, 1952, pp 274-282.

[13] Pimentel, D. and D. Levitan. Pesticides: amounts
     applied and amounts reaching pests. Bioscience, 36,
     1986, pp 86-91.

[14] Gohlich, H. Deposition and penetration of sprays.
     British Crop Protection Council Monograph, 28, 1985,
     pp 173-182.

[15] Bache, D. H. Prediction and analysis of spray
     penetration into plant canopies. British Crop
     Protection Council Monograph. 28, 1985, pp 183-190.

[16] Himel, C. The optimum size for insecticide spray
     droplets. Journal of Economic Entomology, 62, 1969.

[17] Reed, T. F. Advances in spray nozzle designs for
     chemical application. Spraying Systems Co. Wheaton, IL
     p 24.

[18] Gebhardt, M. R. Rotary Disk Atomization. Weed
     Technology, 2, 1988, pp 106-113.

[19] Hislop, E.C. Electrostatic Groun-rig Spraying. Weed
     Technology, 2, 1988, pp 94-105.

[20] Buhler, D. D. and O. C. Burnside. Effect of
     application factors on postemergence phytotoxicity
     of fluazifop-butyl, haloxyfop-methyl and sethoxydim.
     Weed Science, 3, 1989, pp 60-66.

[21] Reichard, D. L. and G. B. Triplett. Paraquat efficacy as influenced by atomizer type. Weed Science, 31, 1983, pp 779-782.

[22] Adams, A. J. and A. Palmer. Deposition patterns of small droplets applied to a tomato crop using the Ulvafan and two prototype electrostatic sprayer. Crop Protection, 5, 1986, pp 358-364.

[23] Frick, E. L. the effect of volume, drop size and concentration, and their interaction, on the control of apple podery mildew by dinocap. British Crop Protection Council Monograph 2, 1970, pp 23-33.

[24] Simmons, H. C. and J. R. Lehtinen. Characteristics of an electrostatically charged air atomized spray for pesticide application, In Pesticide Formulations and Application Systems: Seventh Volume, 1987, ASTM STP 131, pp 113-132.

[25] Johnstone, D. R. Physics and meteorology. In: Pesticide Application: Principles and Practice. Ed. P. T. Haskell. 1985, pp 97-106.

[26] Reichard, D. L., R. D. Fox, R. D. Brazee and F. R. Hall. Air Velocities delivered by Orchard Air Blast Sprayers. Transactions of the American Society of Agricultural Engineers, 22, 1979, pp 69-80.

Christopher M. Riley[1] and Charles J. Wiesner[2]

OFF TARGET PESTICIDE LOSSES RESULTING FROM THE USE OF
AN AIR-ASSISTED ORCHARD SPRAYER

REFERENCE: Riley, C. M. and Wiesner, C. J., "Off-Target
Pesticide Losses Resulting from the Use of an Air-assisted
Orchard Sprayer", Pesticide Formulations and Application
Systems: 10th Volume, ASTM STP 1078, L. E. Bode, J. L.
Hazen, and D. G. Chasin, Eds., American Society for Testing
and Materials, Philadelphia, 1990.

ABSTRACT: This report details the off-target spray losses
resulting from the application of five insecticide sprays to an
orchard using a Good Boy ®air assisted sprayer. Off-target
deposition on the ground was assessed using a number of flat
plate collectors placed at distances up to 200 m downwind of
the orchard. Time integrated airborne spray flux was measured
using Rotorods ® and a novel array of sampling wires. The
results are discussed and used to produce regression equations
from which realistic worst case deposits are predicted.

KEYWORDS: pesticide, drift, spray, orchard, deposition, off-
target.

Airblast sprayers are commonly used to apply pesticides in orchards.
This type of equipment employs large volumes of air to direct and carry spray
droplets up into the tops of fruit trees. Consequently, spray droplets which
do not immediately impact upon the foliage may be carried above and

[1] Pesticide Application Specialist, Department of Chemical and Biotechnical
Services, Research and Productivity Council, P. O. Box 20000, Fredericton,
New Brunswick, Canada, E3B 6C2.

[2] Senior Scientist, Department of Chemical and Biotechnical Services,
Research and Productivity Council, P. O. Box 20000, Fredericton, New
Brunswick, Canada, E3B 6C2.

beyond the relatively sheltered environment of the orchard canopy. Once the airblast has dissipated, movement of smaller droplets with negligible sedimentation velocities is influenced largely by local atmospheric conditions. This may result in a proportion of the emitted material moving out of the intended target area.

A search of the literature revealed that although several studies examined on-target spray deposition resulting from the use of sprayers in an orchard environment [1-6], there was a marked lack of information concerning the resultant off-target drift and deposition of pesticides from orchard spray applications. No information detailing airborne spray cloud concentrations or airborne spray flux from orchard spray applications was found.

The objectives of this study were to assess off-target spray deposition and airborne spray flux resulting from typical orchard spray applications made under a range of atmospheric conditions.

MATERIALS AND METHODS

Site Description and Sampling Scheme

The apple orchard selected for the study was approximately 2 ha in area and located in the St. John River Valley near Lower Burton, New Brunswick. The trees, which were free standing with a maximum canopy diameter of 8 m and a height of 4 to 5 m, were spaced in a grid pattern approximately 10 m apart. Each row of trees was approximately 120 m long and aligned on a heading of 30° east of true north. On the eastern edge of the orchard was located a strawberry field approximately 200 m X 200 m in size. The land sloped gently down from the south to the St. John river in the north. The prevailing wind direction was from a west to south westerly direction.

The sampling equipment consisted of ground deposit samplers, wire arrays, Rotorod samplers and water sensitive cards. A total of twenty-four 20 X 40 cm stainless steel plates (ground deposit samplers) were located at distances of 0, 10, 25, 50, 100 and 200 m downwind of the orchard (Figure 1). At each distance four plates were placed in a line parallel to the edge of the orchard. Sampling plates at the 0 m location were placed along the centre line of the last row in open areas between the trees. The remainder were placed on the soil ridges in open areas between strawberry plants. Plates located at the 10, 25 and 50 m sampling locations were placed at 10 m intervals whereas those at the 100 m and 200 m sampling locations were separated by only 7 m. This was done to reduce the possibility of the drifting spray cloud "missing" one of the sample plates. Only the plates located at the 0, 10, 25 and 50 m sampling distances were in direct alignment. Those at 100 m and 200 m were off-set to accommodate the south westerly wind direction as required. All plates were placed on hardboard in order to reduce

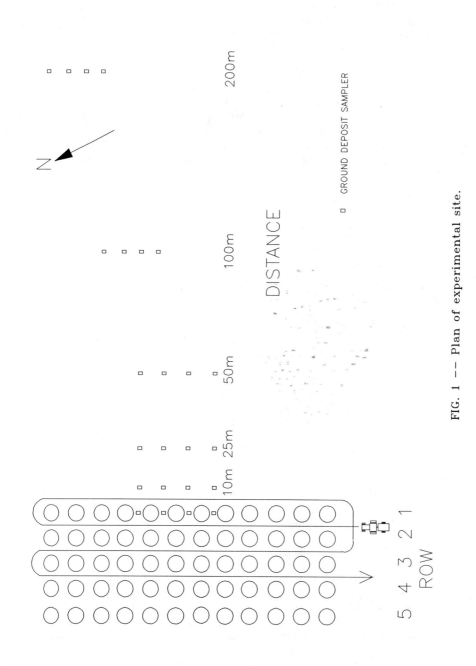

FIG. 1 -- Plan of experimental site.

contamination from dew or soil.

Spray drift flux was assessed by two different methods: (i) "passive" wire samplers and (ii) "active" Rotorod samplers.

Wire arrays were erected at the 50 and 100 m sampling stations. Each array consisted of two 11 m poles between which were strung six horizontal lengths of 0.25 mm diameter monofilament brass wire at heights of 1.75, 3, 5, 7, 9 and 11 m. In an effort to ensure that the entire width of the wire array intercepted the drifting spray cloud, the array at the 100m sampling station was only 10 m wide whereas the array at 50 m was 20 m wide.

A total of 31 Rotorod[3] samplers were deployed downwind of the orchard. Those located at the 25 m and 200 m sample distances were intended to give single measure of time integrated drift flux at "head height" (1.75 m) whereas these located at 50 m and 100 m were intended to provide a means of calibrating the wire samplers and establish a vertical profile of the drift cloud. The location of each sampler is shown in Figure 2.

Water sensitive cards ca 2.5 cm X 10 cm were placed horizontally beside each ground deposit sampler to provide information on the deposited droplet density and an approximate estimate of the deposited droplet spectrum. Cards were also wrapped around the upwind surface of the 13 mm diameter poles on which the Rotorod[®] samplers were supported, or on poles located at each end of the wire arrays. These cards were intended mainly to determine whether the spray cloud had passed by the various sampling surfaces but also provided some information on the drifting spray cloud.

Meteorological Measurements

A 6 m tower was erected and fitted with meteorological instruments to provide information on wind speed and direction at 1.75 m and 6.1 m, air temperature profile to 6.1 m, and relative humidity. Data collected over 10 second execution intervals was averaged over one-minute periods and recorded on a Campbell Scientific 21 X data logger. Atmospheric stability was calculated using ΔT values.[7]

Application Parameters and Trial Procedure

Deltamethrin as Decis®2.5 EC was applied at a nominal rate of 6.0 g a.i./ha as a "low volume" spray in 420 litres of water using a Good Boy® AT1000 p.t.o. driven, air assisted, orchard sprayer. The sprayer was equipped with a 0.71 m diameter fan capable of moving 8.7 m³ air/s at speeds of up to 35 m/s and twelve hydraulic nozzles, six on each side of the sprayer. On each side the upper three nozzles were configured with a ceramic disc having an orifice diameter of 1.2 mm and a swirl plate with a centre orifice diameter

[3]   Ted Brown Associates, 26338 Esperanza Drive, Los Altos Hills, CA 94022

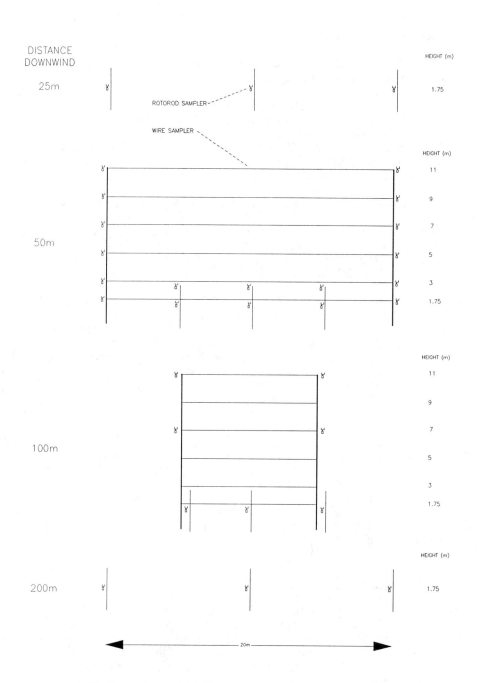

DISTANCE
DOWNWIND

HEIGHT (m)

25m

ROTOROD SAMPLER

1.75

WIRE SAMPLER

HEIGHT (m)

11

9

7

5

50m

3

1.75

HEIGHT (m)

11

9

7

5

100m

3

1.75

HEIGHT (m)

200m

1.75

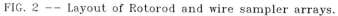

20m

FIG. 2 -- Layout of Rotorod and wire sampler arrays.

of 1.5 mm. The lower nozzles had a 1.2 mm ceramic disc orifice and swirl plate with no centre orifice. When operated at a pressure of 1035 kPa the total calibrated emission rate was 0.42 l/s. The sprayer was driven at a speed of ca 1.2 m/s (4.3 km/h).

The experiment was carried out under the assumption that the spray emitted closest to the downwind edge of the orchard is that which is likely to contribute most to off-target drift and deposition. Therefore, in four of the five applications only the two rows of trees closest to the downwind edge of the orchard were treated. The sprayer was first driven down between rows one and two with all nozzles operating and then driven back up on the downwind edge of row one with only the nozzles on the right side operating i.e. a total application of approximately 63 litres (Figure 1). In the fourth spray application the sprayer was also driven between rows two and three and three and four with all nozzles operating. This application was intended to provide an approximate measure of the contribution to off-target losses resulting from the spraying of rows within the orchard.

Periods of spray emission were timed and the volumes of liquid used were recorded. Samples of tank mix were taken for analysis in order to calculate dosage rates. Spray specific application parameters are given in Table 1.

Sample Collection

When the spray deposits were visibly dry, the water sensitive cards were collected and placed individually in pre-labelled envelopes. Rubber gloves were used at all times when handling the water sensitive cards.

Deposit sample plates, which were hinged in the centre, were folded (exposed side innermost) and placed by sample distance in pre-labelled manila envelopes. Wires were lowered and wound onto glass rods using a cordless drill and placed in labelled screw topped jars. Rotorods were collected and placed in labelled storage racks. To minimize any potential deterioration all samples were stored in the freezer (-11°C) prior to analysis. Samples were analysed within 5 months of spray application.

Sample Analysis

Cards were analysed microscopically to provide information on the deposited droplet density and the deposited droplet size.   The mean deposited droplet density at each sample distance was calculated for both horizontal and vertical cards from the number of droplet stains observed on each card within a circle of 1.77 cm^2. Approximately one hundred stains on each card were sized in 20 um categories.

Deposited droplet size data were pooled by sample distance for both horizontally and vertically orientated cards. A standard spread factor multiplier of 0.5 (calculated from data provided by Spraying Systems Co.) was

TABLE 1 -- Spray specific application parameters

	Spray 1	Spray 2	Spray 3	Spray 4	Spray 5
Date	Aug 22, 1988	Aug 22, 1988	Sept 1, 1988	Sept 1, 1988	Sept 16, 1988
Time of Application	10:06 to 10:11	19:46 to 19:51	09:52 to 09:57	16:58 to 17:08	07:43 to 07:48
Mean temperature (± SD) at 6.1 m (°C)	12.3 ± 0.1	15.8 ± 0.0	15.4 ± 0.1	22.2 ± 0.1	10.2 ± 0.1
Mean relative humidity (± SD) at 6.1 m (%)	83.2 ± 0.1	60.6 ± 0.7	86.3 ± 0.9	47.9 ± 0.4	94.5 ± 0.1
Mean wind speed (± SD) at 6.1 m (m/s)	2.5 ± 0.6	2.3 ± 0.3	3.3 ± 0.2	3.7 ± 0.4	1.6 ± 0.2
Mean wind direction (± SD) at 6.1 m (°true)	275.4 ± 18.5	282.0 ± 5.6	284.9 ± 2.1	290 ± 5.8	264.3 ± 4.9
Mean wind speed (± SD) at 1.75 m (m/s)	2.1 ± 0.5	1.9 ± 0.3	2.8 ± 0.2	3.1 ± 0.3	0.9 ± 0.1
Mean wind direction (± SD) at 1.75 m (°true)	284.6 ± 18.3	292.4 ± 8.8	293.2 ± 4.2	303.4 ± 7.6	237.2 ± 20.3
Atmospheric stability	very unstable	very stable	very unstable	very unstable	very stable

used to estimate deposited droplet sizes from observed stain diameters. Deposited droplet spectra by volume and by number were calculated.

Ground deposit samplers from the 50 m sampling distance were initially processed and analysed individually to give a measure of deposit variability[4] whereas the remainder were bulked and processed by sampling distance. Sample plates were placed and individually separated by paper clips in 25 X 25 cm aluminum pans, immersed in ethyl acetate and washed over a 30 minute period in an orbital shaker at 100 rpm. The pan was covered with aluminum foil to reduce evaporation and prevent splashing. Washings were concentrated under a stream of dry nitrogen to give a final volume of 1 ml to 100 ml depending upon sample location. This processing technique had previously been shown to give good recoveries of deltamethrin [8].

Whilst still wound on the glass rods, wire samples were placed in screw topped jars, immersed in ethyl acetate and washed on an orbital shaker at 200 rpm for 20 minutes. Washings were concentrated to give final volumes of 2 ml and 1 ml from the 50 and 100 m arrays, respectively. This processing technique gave 100% recovery when wires were spiked with 500 ng of deltamethrin.

With the exception of the 50 m Rotorods at the 1.75 m height which were analysed individually, Rotorod samples from each spray were grouped by sample distance and by sample height prior to processing. The rods were placed in wide mouthed screw topped jars containing 50 ml of ethyl acetate and washed with occasional agitation over a 20-30 minute period. The Rotorods were rotated in the jars so that each collection arm in turn was completely immersed in solvent for at least ten minutes. Prior to analysis washings were concentrated to a volume of 1 ml.

One hundred millilitre sub-samples of each tank mix were saturated with an excess amount of NaCl and shaken vigorously with 100 ml of ethyl acetate in a separatory funnel. When the phases had separated, the lower aqueous phase was transferred to a suitable container. The aqueous phase was similarly extracted a second time and the two ethyl acetate phases for each tank mix combined and diluted for GC analysis.

Samples were analysed on a Hewlett Packard 5890 gas chromatograph equipped with a Ni[63] electron capture detector and autosampler. A 10 m X 0.53 mm ID HP-1 Megabore column was used throughout. Helium was used as the carrier gas with a column head pressure of 34 kPa at an oven temperature of 245°C. The makeup gas was 5% methane in argon. Single 1 ul injections of each sample were made with standards injected at frequent intervals.

---

[4]   Having found only trace deposits individual plate washings were combined, concentrated and re-analysed.

RESULTS AND DISCUSSION

None of the spray applications were made at mean wind speeds of less than 0.9 m/s measured at a height of 1.75 m. Local features were such that as wind speed decreased with the development of a stable atmosphere downslope flow caused the low level winds to swing in a more southerly direction and downwind sampling of the orchard was no longer possible.

Water sensitive cards were deployed in only the first four spray applications. Cards were not deployed in the fifth application because of the high humidity conditions prevailing at the time. In each case deposit stains were seen on all cards, even those located at the furthest sampling distance (200 m). Deposit stains in the near field i.e. up to 10 m from the downwind edge of the orchard, were generally smeared. At greater distances stains were approximately circular. Graphs of the deposited droplet densities observed on horizontal and vertical cards are presented in Figures 3 and 4. Results have been standardized to a total application rate of 63 litres to rows one and two as previously described. The results for spray four also include the additional off target deposit occurring as a result of spraying the two extra rows.

Horizontal cards at 0 m were so drenched with liquid that it was not possible to obtain a measure of deposit density. Similarly a large number of deposit stains at the 10 m distance were overlapped and the results which are given represent a lower estimate of the actual deposit density observed. In each case a rapid decline in the deposit density on horizontal cards was seen over the first 50 m with a reduction of at least one hundred times seen over the distance from 10 to 50 m downwind of the orchard. Thereafter deposit density was roughly constant to 200 m. A large proportion (ca 25% to 60%) of all deposit stains at the 100 and 200 m sampling distances were less than 20 um in diameter. In many instances it was difficult to make the distinction between true droplet stains, flaws in the water sensitive cards and small stains caused by atmospheric humidity. The apparent levelling out of deposit density at the further sampling locations may not, therefore, be a true indication of the situation in reality. Similar trends were also seen with the deposition on vertical cards, located from 25 to 200 m downwind of the orchard.

Table 2 details some characteristics of the deposited droplet spectra in sprays one to four. Only deposits from 25 m and beyond were sized because of the problems previously described. As expected a general decrease in $D_{N\ 0.5}$ (Number Median Diameter) was seen with increased distance downwind of the orchard. Similar though less consistent trends were also seen in the value of the $D_{V\ 0.5}$ (Volume Median Diameter). This was due to the fact that the spectra at 100 and 200 m were based on fewer droplet counts and were highly subject to the influence of only a few large droplets.

Quantitative off-target deposition on ground samplers decreased in a log-linear manner with increased distance downwind of the orchard (Figure 5). Results have been standardized to an emitted source strength of 5 mg

TABLE 2 -- Deposited droplet spectra on water sensitive cards

	Spray 1			Spray 2			Spray 3			Spray 4		
	$D_{vo.5}$ (μm)	$D_{No.5}$ (μm)	Dmax category (μm)	$D_{vo.5}$ (μm)	$D_{No.5}$ (μm)	Dmax category (μm)	$D_{vo.5}$ (μm)	$D_{No.5}$ (μm)	Dmax category (μm)	$D_{vo.5}$ (μm)	$D_{No.5}$ (μm)	Dmax category (μm)
Horizontal												
25 m	87	62	150-160	119	71	210-220	129	71	200-210	148	85	250-260
50 m	107	68	140-150	90	10	120-130	111	70	200-210	125	8	130-140
100 m	34	14	40-50	30	14	40-50	142	10	150-160	104	55	210-220
200 m	27	13	60-70	107	12	150-160	17	9	30-40	43	8	70-80
Vertical												
25 m	97	62	160-170	98	63	150-160	115	71	200-210	111	75	200-210
50 m	101	70	160-170	87	65	120-130	94	65	130-140	115	73	240-250
100 m	81	18	110-120	40	11	50-60	85	53	150-160	86	26	110-120
200 m	45	17	60-70	103	13	150-160	59	13	110-120	25	15	40-50

Dmax - size category of largest observed deposited droplet.

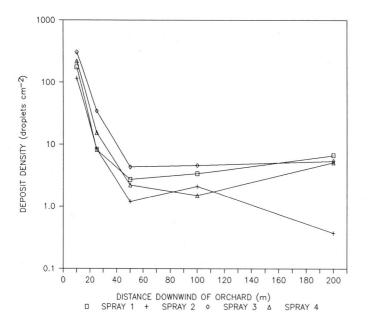

FIG. 3 -- Spray deposit density on horizontal cards.

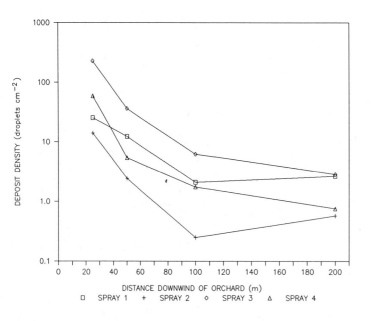

FIG. 4 -- Spray deposit density on vertical cards.

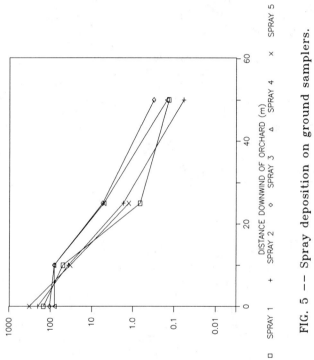

FIG. 5 -- Spray deposition on ground samplers.

deltamethrin/m/row.   In sprays one, two, three and five, therefore, the total standardized amount of active ingredient (a.i.) emitted for each metre of the 120 m long orchard was 7.5 mg i.e. 5 mg/m emitted as the sprayer was driven between rows 1 and 2 and 2.5 mg/m as the sprayer was driven on the downwind side of row 1 with only the right hand nozzles operating.   In the fourth spray the total standardized amount of active ingredient emitted was 17.5 mg/m i.e. 7.5 mg/m as above plus two times 5 mg/m as the sprayer was driven between rows, 2 and 3 and 3 and 4 respectively.

From Figure 5, it can be seen that off-target deposition decreased by a factor of almost one thousand from 0 to 50 m downwind of the orchard. No measurable deposits were detected by GC analysis at either the 100 m or the 200 m sampling stations. The limits of detection and quantification were ca 50 ng/m^2 and 100 ng/m^2 respectively. Deposition on the ground at the 0 m line and the reduction in deposition over the first 10 m were inversely related to wind speed. In sprays three and four (the two highest wind speed sprays) off-target deposition at the 10 m distance was approximately equal to that at the 0 m distance. Off-target deposition in spray four was no greater than that in spray three despite the treatment of two additional swaths. Highly significant correlations between off-target deposit and downwind distance were obtained for each spray application ($P < 0.05$). Calculations are based on the spray deposition observed in the first 100 m downwind of the orchard. Deposits at the 100 m distance were taken to be equivalent to the limit of detection (50 ng/m^2). Correlation co-efficients and regression equations are given in Table 3. Regression line slopes were very similar for all five applications.

The highest off-target deposition is predicted by the regression equation for spray three which was carried out in a wind speed of 3.3 m/s i.e. slightly over the maximum wind speed of 3.1 m/s recommended by the Ontario Department of Agriculture and Food [9]. Cumulative off-target deposition resulting from multiple row applications can be modelled by superimposing, with sequential 10 m offsets, the deposits predicted by this equation. For this purpose calculated off-target spray deposition at distances greater than 90 m downwind of the treated row was assumed to be constant and equal to that predicated for the 100 m distance. Such a calculation clearly represents an overestimate of off-target deposition particularly since the equation was developed from deposit data resulting from off-target movement of spray over essentially bare ground. In a situation involving multiple row treatments off-target movement and deposition of spray released from within the orchard is likely to be reduced by intervening trees. A graph showing the predicted levels of off-target deposition on the ground resulting from the treatment of one, two, three, five and ten rows is presented in Figure 6. Deposit values beyond 100 m have not been predicted because of the uncertainty involved with extrapolating the regression equations too far into the areas where no deposition was detected. The results assume a row spacing of 10 m and an emission rate of 5 mg. a.i. per metre travelled.

Predicted worst case values of off-target deposition on the ground have been integrated with respect to distance to calculate total off-target deposition as a percentage of the material emitted. The results for distances

TABLE 3 -- Regression equations

Spray	Stability	Wind speed at 1.75 m (m/s)	Equation	R^2 value
1	very unstable	2.1	log(deposit) = -0.034 * distance + 1.561	0.77
2	very stable	1.9	log(deposit) = -0.036 * distance + 1.636	0.78
3	very unstable	2.8	log(deposit) = -0.035 * distance + 1.868	0.91
4	very unstable	3.1	log(deposit) = -0.035 * distance + 1.770	0.87
5	very stable	0.9	log(deposit) = -0.037 * distance + 1.661	0.74
combined	--	--	log(deposit) = -0.035 * distance + 1.699	0.80

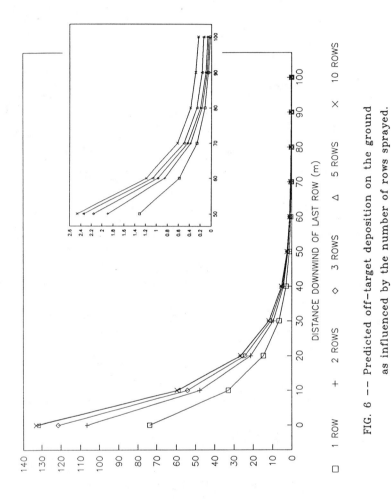

FIG. 6 -- Predicted off-target deposition on the ground as influenced by the number of rows sprayed.

up to 30, 50 and 100 m downwind of the orchard are presented in Table 4. Once again these results reflect the large proportion contributed to drift by the first few rows treated and also the rapid decline in off-target deposition with increased distance downwind of the orchard.

In the only relevant reference found in published literature, MacCollom et al [10] examined the off-target deposition of carbaryl and captan when applied using the Ag-tech and Kinkelder airblast sprayers. Both materials were applied as wettable powder formulations in total volumes of 187 l/ha and 94.6 l/ha depending on the sprayer used. Dosage rates ranged from 1.92 to 3.36 kg a.i./ha. The orchard covered an area of 18 ha and the trees, approximately 6.1 m tall, were on a spacing of 9.1 X 9.1 m. Maximum off-target deposits at 50, 150 and 300 m downwind were 4.17, 0.54 and 0.24 mg/m^2 respectively. No deposits were detected at a distance of 500 m. Unfortunately the dimensions of the orchard were not given in the paper.

TABLE 4 -- Integrated off-target deposition on the ground

Rows treated	Integrated deposit (% of emitted)		
	To 30 m	To 50 m	To 100 m
1	11.7	12.6	12.8
2	10.2	11.0	11.1
3	8.3	8.9	9.1
5	5.7	6.2	6.3
10	3.0	3.3	3.3

Based upon deposit predictions for multiple row treatments in the present study, if deltamethrin were to be applied at 3.36 kg ai/ha the maximum deposit at 50 m would be only 1.68 mg/m^2 or approximately 40% of that observed for captan by MacCollom et al. This difference might be due to differences in emitted droplet spectra, canopy structure, airblast strength or meteorological conditions, the increased drift potential of wettable powder formulations or the evaporation of highly dilute emulsifiable concentrate sprays resulting in droplet sizes having a tendency to remain airborne rather than to be deposited on the ground.

With the exception of spray five, deposits were observed on all wire arrays at the 50 m distance, however, in only two cases, spray two and spray four were quantifiable deposits obtained on the 100 m sampling array (Table 5). Using the equation given by Edmonds [11] it can be shown that the 0.25 mm diameter sampling wires theoretically have a very high capture efficiency for small droplets (> 90% for droplets of 50 μm diameter at a wind speed of 1 m/s). Therefore, if it is assumed that the wires intercept all the droplets which would have passed through the airspace which they occupy, a quantitative value of time integrated spray flux can be calculated based upon the total deposit observed. (Spray flux is the amout of material passing through an area in the vertical plane, the horizonal axis of which is parallel to, and downwind of the edge of the target area. The units of flux are $g/m^2/s$. Time integrated flux is expressed as g/m^2).

From the values in Table 5, it is apparent that in the first four sprays the drifting spray cloud extended above the 11 m sampling array located at the 50 m sampling station. The highest values, equivalent to 4.7 ug/m^2 were seen in spray four, the highest wind speed spray in which four rows were treated. The time integrated spray flux values can also be integrated vertically and expressed as a percentage of the total material applied to rows one and two i.e. 7.5 ug. These values, given in Table 6, show that below a height of 11 m less than 0.5% of the material applied was still airborne at a distance of 50 m from the edge of the orchard.

Treatment of the additional rows in spray four produced an increase in off-target drift at the 50 m distance when compared to the other spray applications. This increase was equivalent to approximately 30% when compared to the very stable application at 1.9 m/s and approximately 70% when compared to the very stable application at 2.8 m/s. Integrated flux and hence mean drift cloud concentration decreased with increasing distance downwind of the orchard. This is indicative of material losses due to droplet deposition and of vertical and horizontal diffusion of the drift cloud. In instances where no deposits were detected on the wires at a distance of 100 m, this is probably because the deposits were below the detection limit since deposits were observed on water sensitive cards placed at each end of the arrays. Water sensitive cards were not used in spray five and it was not, therefore, possible to verify whether or not the sampling arrays were actually deployed within the path of the drifting spray cloud.

With the exception of three Rotorod samples at the 50 m sampling distance in which trace amounts of deltamethrin were found (one at a height of 3 m in spray one and two at heights of 3 m and 5 m in spray four) the only measurable deposits were found on the Rotorod samplers located at the 25 m sampling distance. The detection limit for deltamethrin in the Rotorod sample washings was ca 0.005 ug/ml irrespective of the number of samplers from which the sample was obtained. This value can be used to calculate the upper limit of flux values for samples on which no deposit was observed. Time integrated spray flux values for samplers located at the 25 m sampling distance and upper limit time integrated flux values, integrated to a height of 11 m for samplers located at the 50 m sampling distance are presented in Table 7.

Time integrated flux values in sprays one, two and five were very similar at the 25 m sampling distance and were highest in the fourth spray application. Once again, the amount of material still airborne below a height of 11 m was calculated to be less than 0.5% at the 50 m sampling distance. With the exception of spray two, integrated flux values calculated from observed deposits on wire sampling arrays were less than the maximum

TABLE 5 -- Spray deposition on wire arrays

Spray	Sample height (m)	50 m Sampling Distance		100 m Sampling Distance	
		Total deposit (ug)	Time integrated spray Flux (ug/m^2)	Total deposit (ug)	Time integrated spray Flux (ug/m^2)
1	1.75	ND	--	ND	--
	3	0.008	1.5	ND	--
	5	ND	--	ND	--
	7	0.008	1.5	ND	--
	9	0.011	2.3	T	0.8
	11	0.008	1.5	ND	--
2	1.75	0.008	1.7	0.003	1.0
	3	0.008	1.7	0.003	1.2
	5	0.009	1.8	0.004	1.7
	7	0.010	2.0	0.003	1.3
	9	0.008	1.7	0.003	1.3
	11	0.006	1.2	0.003	1.3
3	1.75	0.008	1.5	ND	--
	3	0.008	1.5	ND	--
	5	0.005	0.9	ND	--
	7	0.008	1.5	ND	--
	9	0.006	1.2	ND	--
	11	0.005	0.9	ND	--
4	1.75	0.024	4.7	ND	--
	3	0.017	3.5	0.004	1.6
	5	ND	--	0.009	3.5
	7	ND	--	T	0.9
	9	0.013	2.5	T	0.9
	11	0.011	2.2	0.004	1.6
5	1	ND	--	ND	--
	3	ND	--	ND	--
	5	ND	--	ND	--
	7	ND	--	ND	--
	9	ND	--	ND	--
	11	ND	--	ND	--

ND - not detected.   T - trace.
Limit of detection ca 2 ng.   Limit of quantification ca 3 ng.
Results standardized to 5.0 mg a.i./metre/row (see text for details).

TABLE 6 -- Time integrated off-target spray flux to a height of 11 m as calculated from spray deposits on wire arrays. (% of emitted)

Spray	50 m sampling station	100 m sampling station
1	0.15	ND
2	0.25	0.19
3	0.19	ND
4	0.32	0.17
5	ND	ND

ND - not detected (< 0.06% at 50, < 0.12% at 100 m).

TABLE 7 -- Time integrated spray flux as measured by Rotorod samplers

Spray	25m sampling station (ug/m^2)	Integrated to a height of 11 m at 50 m sampling distance (% of emitted)
1	6.98	0.29
2	6.90	0.11
3	16.09	0.31
4	24.40	0.35
5	7.01	0.10

[a]   Standardized to an emission rate of 7.5 mg a.i./m.

values calculated indirectly from the lack of measurable deposits on Rotorod samplers. Due to the lack of data points with quantifiable deposits no attempt was made to calculate regression equations relating integrated spray flux with distance downwind of the orchard.

CONCLUSIONS

Off-target spray deposition decreased rapidly with increased distance downwind of the orchard. In each of the five applications deposition on ground samplers decreased by approximately three orders of magnitude over a distance of 0 m to 50 m. No deposits were detected by GC analysis at distance of 100 m and 200 m. Decreases in quantitative deposit were accompanied by similar reductions in deposit density as well as mean droplet size.

Regression equations of deposit vs downwind distance were used to predict cumulative, worst case deposits that might realistically be expected from multiple row applications. The results indicate that spray applications made to the first five rows, closest to the downwind edge of the orchard, cumulatively make up more than 99% of off-target deposits on the ground. Predicted deposits from multiple row applications at downwind distances of 10 m, 50 m, and 100 m were equivalent to approximately 11.9%, 0.5% and 0.05% of the nominal dosage rate respectively.

When only the two rows closest to the downwind edge of the orchard were sprayed, the maximum value of time integrated airborne spray flux integrated to a height of 11 m at the 50 m downwind distance was equivalent to 0.32% of the emitted material. However, vertical deposit profiles indicated that the drifting spray cloud extended above the maximum 11 m sampling height. Similarly, maximum time and height integrated flux observed at a distance of 100 m was equivalent to 0.19% of emitted material.

Whilst it is obviously impossible to test every application scenario it should be remembered that off-target pesticide losses are likely to be influenced not only by meteorological conditions but also the characteristics of sprayer used (e.g. airblast strength, droplet size, nozzle orientation and arrangement), the type of formulation used (e.g. emulsifiable concentrate vs wettable powder), and tree morphology (e.g. height, foliar development and canopy density). Although this study is one of the very few which addresses off-target pesticide losses from orchard spray operations it deals with only a limited range of application variables. Consequently, the results should be considered only as preliminary guidelines when evaluating the potential effects of off-target pesticide losses on public health.

ACKNOWLEDGMENTS

This study was instigated and funded by the New Brunswick Department of Health and Community Services. The authors wish to thank Mr. S. Rosenfeld and Mr. R. E. Crossman for their assistance and Dr. E.

Estabrooks and Agriculture Canada for the use of the sprayer employed in this study.

REFERENCES

[1]   Carman, G. E., Iwata, Y. and Gunther, F. A. "Pesticide Deposition on Citrus Orchard Soil Resulting from Spray Drift and Run off". Bulletin of Environmental Contaimination and Toxicology. Vol. 18, No.6., 1977, pp 706-710.

[2]   Hall, F. R., Ferree, D. C., Reichard, D. L. and Krueger, H. R. "Orchard Geometry ad Pesticide Deposition Efficiency". In "Fruit Crops 1987: A Summary of Research". Ohio Agricultural Research and Development Center Research Circular 295, Wooster, Ohio. July 1988, pp 23-27.

[3]   Inculet, I. I., Castle, G. S. P., Menzies, D. R. and Frank, R. "Deposition Studies With a novel form of Electrostatic Crop Sprayer". Proceedings of the fourth International Conference on Electrostatics, The Hague, The Netherlands, May 1981.

[4]   Reichard, D. L., Hall, F. R. and Retzer, H. J. "Distribution of Droplet Sizes Delivered by Orchard Air Sprayers". Journal of Economic Entomology. Vol. 71, No. 1, 1978, pp 53-57.

[5]   Reichard, D. L., Hall, F. R. and Krueger, H. R.   "Effects of Application Equipment Variables on Spray Deposition by Orchard Air Sprayers".   Ohio Agricultural Research and Development Center. Research Circular 272, Wooster, Ohio, August 1980.

[6]   Van Ee, G. R., Ledebuhr, R. L. and Potter, H. S. "Development of a CDA/Air Carrier Sprayer". American Society of Agricultural Engineers. Paper #84.1507.

[7]   Sedefian, L. and Bennett, E.   "A Comparison of Turbulence Classification Schemes". Atmospheric Environment. Vol. 14, 1980, pp. 741-750.

[8]   Riley, C. M., Wiesner, C. J. and Ernst, W. R. "Off-target deposition and drift of aerially applied agricultural sprays", Pesticide Science. Vol. 26, No. 2, June 1989, pp. 159-166.

[9]   Fisher, R. W. and Hikichi, A., Orchard Sprayers. Ontario Department of Agriculture and Food. Publication 373. Toronto, Ontario, 1988.

[10]   MacCollom, G. G., Currier, W. W., and Baumann, G. L.   "Drift Comparisons between Aerial and Ground Orchard Application". Journal of Economic Entomology. Vol. 79, No. 2, April 1986, pp. 459-464.

[11]   Edmonds, R. L.   "Collection Efficiency of Rotorod Samplers for Sampling Fungus Spores in the Atmosphere". Plant Disease Reporter. Vol. 56, No. 8, August 1972, pp. 704-708.

Palaniappa Krishnan, John Herrera, L. James Kemble, and Stephan Gottfried[1]

EFFECT OF SPRAYER BOUNCE AND WIND CONDITION ON SPRAY PATTERN DISPLACEMENT OF TWO AGRICULTURAL NOZZLES

REFERENCE: Krishnan, P., Herrera, J., Kemble, L. J., and Gottfried, S., "Effect of Sprayer Bounce and Wind Condition on Spray Pattern Displacement of Two Agricultural Nozzles," Pesticide Formulations and Application Systems: 10th Volume, ASTM STP 1078, L. E. Bode, J. L. Hazen, and D. G. Chasin, Eds., American Society for Testing and Materials, Philadelphia, 1990.

ABSTRACT: Effect of sprayer bounce and wind velocity on spray pattern displacement (SPD) of two agricultural nozzles were studied using an experimental spray patternator. Tests were conducted at a nozzle pressure of 207 kPa at two wind conditions (C, crosswind velocity of 2.68 m/s; and CH, combination of crosswind velocity and headwind velocity of 2.63 m/s) and at two sprayer bounces (0.2 and 0.4 m amplitude at a frequency of 1 Hz).

Sprayer bounce x wind velocity had significant effect on the SPD values. For both the nozzles, the SPD value for sprayer bounce (0.4 m amplitude) x crosswind velocity x headwind velocity was higher by at least 5 percentage points than that for sprayer bounce (0.4 m amplitude) x crosswind.

KEYWORDS: spray patternator, crosswind, headwind, coefficient of variation, spray boom movement.

Two popular fan nozzles are the 8004VS and the 11004VS fan nozzles manufactured by Spraying Systems Co., N. Avenue, Wheaton, IL 60138. These nozzles are commonly used for applying pre-emerge surface applied herbicides and post-emerge contact herbicides. They produce a tapered edge, fan spray pattern.

Recommendations for equipment operating conditions for obtaining uniform deposits are primarily based on laboratory data, and, in most cases, are not compatible with field operations [1]. This is true because laboratory tests have been conducted under field tests have always been under dynamic conditions in presence of wind and sprayer bounce.

---

[1]Dr. Krishnan is assistant professor, Mr. Herrera is an undergraduate student, Mr. Kemble and Mr. Gottfried are research associates in the Agricultural Engineering Department, University of Delaware, Newark, DE   19717-1303.

Krishnan et al. [2] and Krishnan [3] have developed a technique for measuring spray pattern displacement (SPD's) of agricultural nozzles using a spray table operated under static conditions (absence of wind and sprayer bounce), whereas the dynamic conditions. A minimal SPD value would ensure an effective chemical application.

A zero value for the SPD component would indicate an ideal chemical spray application. Krishnan et al. [4] studied the effects of nozzle pressure and wind velocity on SPD for 8004 and XR8004 fan nozzles (Spraying Systems Co.). However, the effect of sprayer bounce was not studied at that time.

Speelman and Jansen [5] found that the vibrations of the spray-boom of field crop sprayer affected the liquid distribution in a negative way.   They studied the intensity and character of the vibrations of boom ends of four tractor-mounted field crop sprayers in the horizontal and the vertical planes.

There has been very little published work on studies of agricultural nozzles under dynamic conditions in the laboratory. The objective of this study was to study the effect of sprayer bounce and wind velocity on spray pattern displacement (SPD) of two agricultural nozzles.

EXPERIMENTAL METHOD

Preliminary Field Tests

Preliminary field tests were carried out to determine the bounce frequency and amplitude for designing the spray bounce simulator in the laboratory.

The equipment included a JD2350 tractor, an experimental 303-liter mounted sprayer, and nozzles. The sprayer was half filled. A piece of angle iron was attached to the sprayer frame (Fig. 1). The motion of the angle iron was independent of the motion of the boom. A counter trigger was installed at the other end of the angle iron. It was activated by a rubber band connected to the end of the boom. This counter trigger was used to record the vertical frequency of the boom.   A marker board was attached to the boom.   A thick felt tip marker was attached to the end of the angle iron. As the sprayer traversed the field, the felt tip marker moved against the spring pressure in a vertical plane making marks on the board. The number of peaks made by the marker was read by the counter trigger and the time taken to tranverse the given distance was noted.   The vertical frequency of oscillation in Hz of the boom was computed from these measurements.

A 61-m course was marked on an alfalfa field. The tractor field speed was 10-11 km/h. A low gear setting with a high rpm was used to obtain the desired field speed quickly. Four field trials were carried out and the vertical amplitude of the boom and the boom frequency was measured. The trials were videotaped. The tape was played and the peaks made by the marker were counted. Knowing the time taken to traverse the 61-m course, the frequency of oscillation of the

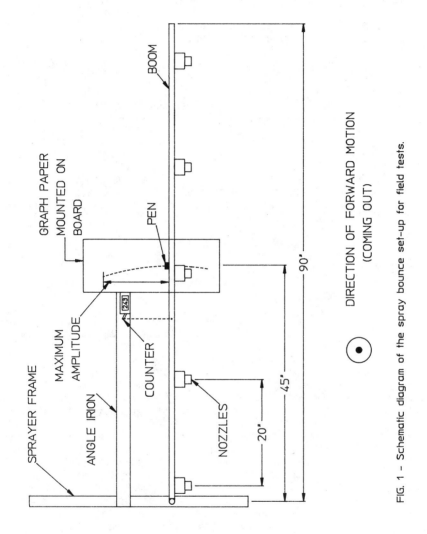

FIG. 1 – Schematic diagram of the spray bounce set-up for field tests.

boom was computed. These values were in good agreement with the values obtained from counter trigger measurements.

Laboratory Tests

Equipment Description: An experimental aluminum spray patternator (4.27 m long and 2.44 m wide) constructed with 111 right angled grooves measuring 38.4 mm peak-to-peak were used. A detailed description of the equipment can be found in Krishnan et al. [2]. A 0.91 m axial fan (Chore Time, model 36 pf, 0.4 kW, 530 rpm) was used to simulate the crosswind. Two 0.61 m axial fans (Dayton, model 4c 354c, 0.2 kW, 1075/2 speed, 3.5 A, 115 V, 1 phase) were used to simulate the headwind. Shrouds with baffles were constructed for these fans (Fig. 2). The wind speed was regulated by adjusting the baffles and by controlling the air flow through the fan.

Test Parameters: Tests were conducted with five 8004VS and five 11004VS fan nozzles manufactured by Spraying Systems Co., N. Avenue, Wheaton, IL 60188. The nozzle spacing was kept constant at 0.51 m. The nozzles were selected at random for the placement on the boom. The target width was 1.02 m. Tests were conducted at a nozzle pressure of 207 kPa at two wind conditions (C, crosswind velocity of 2.68 m/s; and CH, combination of crosswind velocity and headwind velocity of 2.63 m/s).

The wind velocity was measured using a TurboMeter$_{TM}$ wind speed indicator. The crosswind velocity was measured at 27 points parallel to the line of the boom and then averaged. The headwind velocity was measured at nine points on a plane containing the three nozzles and then averaged.

Tests were conducted at two sprayer bounces (0.2 and 0.4 m amplitude) at a frequency of one Hz. Each test was videotaped.

The optimal height of the nozzle above the spray patternator (OHONASP) that gave a uniform spray distribution was determined for each of the nozzle pressures. Pattern uniformity was determined qualitatively by viewing the videotaped spray patterns and quantitatively by calculating the coefficient of variation (CV) of the spray pattern. A CV of 10% was taken to indicate acceptable coverage because even with new nozzles a C.V. of 10% could be expected. The static tests (STC) and dynamic tests (DTC) were carried out at the corresponding predetermined optimal height for each nozzle pressure.

Before each test run the nozzle pressure was regulated by a throttling valve at the pump. Some time was allowed for the pump pressure and the nozzle flow to reach stable levels. When the desired pressure was obtained at the nozzle, the graduated cylinders were properly positioned beneath each groove and water was collected for 60 s.

Spray Pattern Displacement Measurements: The spray pattern displacement (SPD) measurements were made using equation 1.

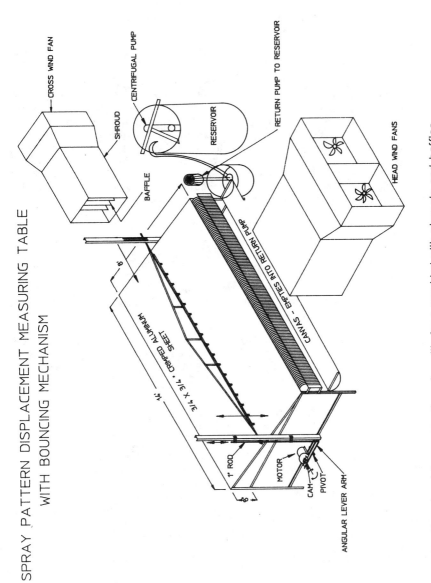

SPRAY PATTERN DISPLACEMENT MEASURING TABLE WITH BOUNCING MECHANISM

FIG. 2 – Experimental spray patternator with fans provided with shrouds and baffles.

$$SPD = \frac{\displaystyle\sum_{n=1}^{N} \left| (Vnw(n) - Vw(n)) \right|}{\displaystyle\sum_{n=1}^{N} Vnw(n)} \times 100 \qquad (1)$$

where: SPD =   spray pattern displacement, %,
                N =   number of graduated cylinders positioned beneath the grooves of the spray patternator within the target width,
      Vnw(n) =   volume of the nth graduated cylinder under no-wind-test condition (STC) within the target width,
       Vw(n) =   volume of the nth graduated cylinder under wind-test condition (DTC) within the target width.

Statistics

An analysis of variance was carried out to test the effect of the wind velocity and sprayer bounce on the SPD values. Duncan's Multiple Range Test at the 0.05 level of significance, using SAS statistical package [6], was used to separate the significant means.

Results

Preliminary Field Tests: Preliminary field tests revealed that the range of the bounce amplitude in the field was 0.1 to 0.2 m and the range of the bounce frequency was 0.08 to 0.1 Hz. These values are reasonable for a fully mounted sprayer. However, for semi-mounted sprayers and trailed sprayers these values may be very different because of their configuration. Further research should be conducted on semi-mounted and trailed sprayers. On both types the distance from the hitch point to the spray boom in relation to the sprayer's axle will have effect on the boom's vertical movement due to the pivot (axle) location. This movement of rotation around the sprayer axle plus the soil surface roughness may create quite different frequencies than experienced with a fully mounted sprayer.

Optimal heights of the nozzle for each test condition: The optimal heights of the nozzle above the spray patternator, OHONASP, at 207 kPa nozzle pressure for the static test condition (STC) for the 8004- and 11004 VS fan nozzles were 0.5 and 0.3 m, respectively. As the dynamic tests, DTCs, were done at the same corresponding heights, the spray distribution pattern was no longer uniform. This provided data to determine the spray pattern displacements (see equation 1 and Tables 1 and 2) for the two nozzles at different DTCs.

Spray Pattern Displacement (SPD) Measurements: The SPD value determines if adequate coverage (total quantity of spray) is provided within the target width. A lower SPD value implies good coverage.

TABLE 1 -- Mean spray pattern displacement[a] (SPD) and coefficient of variation (CV) data for 8004VS fan nozzle at selected nozzle pressure, wind conditions and sprayer bounces.

Nozzle Pressure, kPa	Volume[b] Median Diameter, μm	Treatment[c] Conditions	SPD[d], %	CV, %
207	450	B1	22.4 ± 0.7 CD	9.6 ± 0.9
		B1C	23.4 ± 1.3 BC	9.8 ± 0.7
		B1CH	25.2 ± 1.6 AB	11.9 ± 0.9
		B2	20.1 ± 1.9 D	10.4 ± 1.0
		B2C	20.5 ± 0.9 D	8.6 ± 0.8
		B2CH	26.1 ± 0.5 A	8.5 ± 0.9

[a]Mean SPD and CV data is the average of three replicates. The nozzle spacing was held constant at 0.5 m.

[b]The volume median diameter ($D_{v0.5}$) value was obtained from Spraying Systems Co., North Avenue, Wheaton, IL 60188.

[c]Treatment Conditions:

B1 - Spray bounce of 0.2 m amplitude at 1 Hz
B2 - Spray bounce of 0.4 m amplitude at 1 Hz
C - Crosswind of 2.5 m/s
H - Headwind of 2.2 m/s
B1C - Spray bounce of 0.2 m amplitude x crosswind
B1CH - Spray bounce of 0.2 m amplitude x crosswind x headwind
B2C - Spray bounce of 0.4 m amplitude x crosswind
B2CH - Spray bounce of 0.4 m amplitude x crosswind x headwind

[d]Values followed by the same letter in the SPD column are not significantly different from each other at the 5% level of probability based on Duncan's multiple range test.

TABLE 2 -- Mean spray pattern displacement[a] (SPD) and coefficient of variation (CV) data for 11004VS fan nozzle at selected nozzle pressure, wind conditions and sprayer bounces.

Nozzle Pressure, kPa	Volume[b] Median Diameter, μm	Treatment[c] Conditions	SPD[d], %	CV, %
207	400	B1	23.5 ± 1.5 C	6.9 ± 0.9
		B1C	26.2 ± 2.8 BC	13.0 ± 0.9
		B1CH	26.0 ± 0.5 BC	7.2 ± 0.3
		B2	25.1 ± 1.8 BC	8.9 ± 1.2
		B2C	27.4 ± 0.9 B	11.6 ± 0.8
		B2CH	33.5 ± 2.1 A	8.2 ± 0.8

[a]Mean SPD and CV data is the average of three replicates. The nozzle spacing was held constant at 0.5 m.

[b]The volume median diameter ($D_{Vo.5}$) value was obtained from Spraying Systems Co., North Avenue, Wheaton, IL 60188.

[c]Treatment Conditions:

B1 - Spray bounce of 0.2 m amplitude at 1 Hz
B2 - Spray bounce of 0.4 m amplitude at 1 Hz
C - Crosswind of 2.6 m/s
H - Headwind of 2.2 m/s
B1C - Spray bounce of 0.2 m amplitude x crosswind
B1CH - Spray bounce of 0.2 m amplitude x crosswind x headwind
B2C - Spray bounce of 0.4 m amplitude x crosswind
B2CH - Spray bounce of 0.4 m amplitude x crosswind x headwind

[d]Values followed by the same letter in the SPD column are not significantly different from each other at the 5% level of probability based on Duncan's multiple range test.

The mean SPD values for the 8004VS fan nozzle was 4 percentage points lower (P<0.05) than that for the 11004VS fan nozzle for the treatments tested. There was no significant difference (P>0.05) in the effect the sprayer bounce had on the SPD value for the range tested. However, for the 8004VS fan nozzle, the SPD values obtained in these tests with sprayer bounces were at least 10 percentage points higher than obtained previously under no bounce condition [4]. Similar trend could be expected for the 11004VS fan nozzle also. Sprayer bounce and wind interaction had significant effect on the SPD values. For both the nozzles, the SPD value for sprayer bounce (0.4 m amplitude) x crosswind x headwind was higher by at least 5 percentage points (P<0.05) than that for sprayer bounce (0.4 m amplitude) x crosswind.

In general, the SPD value increased with the CV value. However, for the treatment condition that included the headwind, the SPD value was the highest even though the CV values were at the lower end of the range. The headwind tended only to improve the CV value and not the SPD value of the spray pattern.

Conclusions

1. The mean SPD values for the 8004VS fan nozzle was 4 percentage points lower (P<0.05) than for the 11004VS fan nozzle for the treatments tested.
2. There was no significant difference (P>0.05) in the effect the sprayer bounce had on the SPD value at a bounce frequency of 1 Hz for the nozzles tested.
3. For both the nozzles, the SPD value for sprayer bounce (0.4 m amplitude) x crosswind x headwind was higher (P<0.05) by at least 5 percentage points than for sprayer bounce (0.4 m aplitude) x crosswind.

ACKNOWLEDGEMENTS

The authors acknowledge with appreciation the help of Nancy Cassidy and Fran Mullen in the preparation of this paper.

REFERENCES

[1]   Smith, D. B., "Evaluation of Broadcast Spray Deposits," ASAE Paper No. 83-1512, American Society of Agricultural Engineers, St. Joseph, MI 49085, 1983.

[2]   Krishnan, P., T. H. Williams and L. J. Kemble, "Spray Pattern Displacement Measurements of 8004 and XR8004 Fan Nozzles Using Spray Table," Transactions of the American Society of Agricultural Engineers, Vol. 31, No. 2, 1988, pp 386-389.

[3]   Krishnan, P., "Spray Displacements of Two Agricultural Nozzles Using Spray Patternator," 9th Symposium on Pesticide Formulations and Application Systems "International Aspects," ASTM STP 1036, James Hazen and David A. Hovde, editors, American Society for Testing and Materials, Philadelphia, 1989, pp 254-261.

[4]   Krishnan, P., T. H. Williams and L. J. Kemble, "Spray Pattern
       Displacement Measurements of 8004 and XR8004 Fan Nozzles Using
       Spray Table," Transactions of the American Society of
       Agricultural Engineers, Vol. 31, No. 6, 1988, pp 1660-1663.
[5]   Speelman, L., and J. W. Jansen, "The Effect of Spray-boom
       Movement on the Liquid Distribution of Field Crop Sprayers,"
       J. Agric. Engrg. Res., Vol. 19, 1974, pp 117-129.
[6]   SAS[R] User's Guide: Statistics, Version 5 Edition, SAS
       Institute, Inc., Box 8000, Cary, NC 27511-8000, 1985.

H. Erdal Ozkan, Donald L. Reichard, Harry D. Niemczyk, Michael G. Klein, and Harvey R. Krueger

SUBSURFACE INJECTION OF TURFGRASS INSECTICIDES

REFERENCE: Ozkan, H.E., Reichard, D.L., Niemczyk, H.D., Klein, M.G., and Krueger, H.R., "Subsurface Injection of Turfgrass Insecticides," Pesticide Formulations and Application Systems: 10th Volume, ASTM STP 1078, L.E. Bode, J.L. Hazen, and D. G. Chasin, Eds., American Society for Testing Materials, Philadelphia, 1990.

ABSTRACT: Surface application of insecticides on turfgrass for control of many subsurface grubs is not very efficient. It can be potentially hazardous to people, pets and other animals exposed to the insecticides. To reduce this danger, three injection systems were developed to place insecticides below the turf surface and into the main zone of activity of the grubs. The results indicate that insecticides can be successfully injected with little damage to the turf. Analyses of surface residues after injection of two common turfgrass insecticides showed significant reductions ranging from 38 percent to 95 percent over the amount remaining from spraying on the surface. Grub control using two of the three injection methods was equal to or better than that achieved with the conventional surface application method. Although it satisfactorily injected the insecticide, the third injection method using a rolling point applicator was not evaluated for grub control because the preliminary data showed little dispersion of insecticide from the point of injection and indicated a low probability of achieving efficient and effective control.

KEYWORDS: injection, turfgrass, grub control, insecticides

Dr. Ozkan is Associate Professor of Agricultural Engineering at Ohio State University; Mr. Reichard is Agricultural Engineer, USDA-ARS, Wooster, Ohio; Dr. Niemczyk is Professor of Entomology at Ohio State University; Dr.Klein is Entomologist, USDA-ARS, Wooster, Ohio; Dr. Krueger is Professor of Entomology at Ohio State University.

INTRODUCTION

Pesticides are expected to remain the primary means of controlling various pests of turfgrass. Among the pesticides commonly applied to turfgrass, insecticides and nematicides are generally the most toxic. Currently, some of these pesticides are broadcast on the turfgrass surface for control of white grubs (beetle larvae) and plant parasitic nematodes. Many of the problems associated with the application of soil insecticides would be reduced if the insecticides were applied directly to the primary location of the grubs, the soil-thatch interface. Control of subsurface insects could potentially be achieved with greatly reduced rates since the insecticide would not be absorbed by the thatch before reaching the insects as often occurs with surface applications. Risks of human exposure to insecticides on golf courses, home lawns and other recreational turfgrass areas could potentially also be reduced by subsurface placement. Subsurface application could also reduce run-off, water pollution, hazards to wild life, and would eliminate spray drift.

Numerous studies have been reported and reviews written [1, 2, 3] regarding the behavior of insecticides in soils. Information from studies on the movement and half-life of pesticides required by the Environmental Protection Agency (EPA) for registration, provides an understanding of pesticide movement and dissipation following application on the soil surface. However, insecticides for control of insect pests of turfgrasses are usually not applied directly to soil, but instead to the foliage and a layer of thatch above the soil. Subsequent movement and fate of pesticides for such applications is distinctly different from application directly in or on soil. For insecticide applications on the turfgrass surface, researchers reported the persistence and movement of isazofos, chlorpyrifos, diazinon and chlordane [4]; and studied chlorpyrifos and diazinon following application to turfgrass [5]. Both studies were conducted on turfgrass with little (<0.6 cm) (<0.25 in) or no thatch. Depending on the location, thatch depth can range from 0.6 to 2.5 cm (0.25 to 1 in) and several research studies [6, 7, 8] showed that thatch can adsorb the majority of insecticide applied. Partly because of this reason, recommended application rates of insecticides for grub control are usually high. Recent hearings sponsored by the EPA determined that surface applications of diazinon were responsible for the death of Brandt geese on a Long Island golf course [9]. As a result, the EPA cancelled registration for diazinon use on golf courses and sod farms. The potential for human and wildlife exposure to insecticides on golf courses, home lawns and recreational turf indicates a need to develop an alternative method to surface applications of insecticides. In view of current concerns by the EPA and the public perception about pesticides applied to turfgrasses as possibly contributing to ground water contamination, additional research on efficient application of pesticides is essential.

Machines have been developed to inject liquid fertilizers with little disturbances of the surface. Liquid Ed Inc.[1], Lake Worth, Florida developed a machine to inject fertilizer below plastic film in vegetable crops. Researchers at Iowa State University have developed a similar point injector applicator for application of liquid N, P, K and Anhydrous Ammonia. Their research determined that point-injection provided higher corn yields than surface applications of fertilizers. Other advantages of using the point injector fertilizer applicator included better plant-use efficiency; little loss of fertilizers due to immobilization at the surface, NH_3 volatilization and P and K stratification; and little or no root pruning or soil disturbance with close placement to growing plants [10, 11].

There is considerable need to improve the efficacy and safety of using pesticides for controlling white grubs (beetle larvae) in turfgrasses. The objective of this research was to develop and evaluate equipment for injecting insecticides in turfgrass.

EQUIPMENT AND RESULTS

Point Injection

First, a point injection system using a spoked injection wheel (Fig. 1) was assembled. The self-propelled applicator had a 3.7 kW (5 Hp) engine to propel the applicator and drive a rotary vane pump capable of delivering the liquid at 1380 kPa (200 psi) to the injector. Two injector wheels, one developed at Iowa State University [12], the other one manufactured by Liquid Ed Inc., Lake Worth, Florida, (Figure 2) were modified to inject chlorpyrifos (Dursban) at the thatch-soil interface. Both wheels are similar in design. The axle of the wheel contains a rotary valve which was in the open position only for the spoke that was vertical and into the downward position. In an attempt to improve the uniformity of insecticide distribution surrounding the point of injection, the number of orifices at the end of the spoke was increased from two to four. The orifices were 90 degrees apart from each other. Detailed description of both wheels and the modifications made were reported by the authors in another publication [13].

Several tests were performed with the modified Iowa State wheel. Even with 1000 kPa (145 psi) pressure, some of the orifices at the leading edge and some of the side orifices plugged with soil after brief use in the field. Among the alternative configurations tested, the one with orifices 120 degrees apart showed no evidence of plugging when liquid pressure was 1000 kPa (145 psi).

---

[1]Mention of a trade name or proprietary product is for specific information only and does not constitute a guarantee or warranty of the product by the U.S. Dept. of Agric. or The Ohio State University and does not imply endorsement of the product over other products not mentioned.

Figure 1   Self propelled, walk behind point injector applicator

Figure 2   The point injector wheel manufactured by Liquid Ed Inc., Florida

The orifice on the trailing edge of spoke plugged when liquid pressure was 276 kPa (40 psi).   After a satisfactory combination of design and operating conditions (orifice configuration, pressure, travel speed, etc.) was established, the wheel was tested to determine placement of the liquid in the turf.   A fluorescent tracer and water mixture was injected and several core samples (7.6 cm in diameter and 5.1 cm deep) from the thatch and soil at points of injection were visually inspected under ultraviolet light. The visual inspection of samples indicated that liquid deposits on the surface were unexpectedly high with the Iowa State wheel. Leakage around the hub was also a problem when the wheel was operated with 1000 kPa (145 psi) pressure.   The wheel manufactured by Liquid Ed Inc. did not leak at 1000 kPa (145 psi) pressure and its initial performance was found acceptable. This wheel was calibrated to inject chlorpyrifos into turf at the desired rate and used for all subsequent data.

Four field plots, each 15 m (50 ft) long and 6 m (20 ft) wide, were established on a golf course fairway at Wooster, Ohio for use in determining insecticide residues around the points of injection.   The injector wheel delivered 7.5 ml/injection when operated at 1000 kPa (145 psi) while traveling from south to north at 3.2 km/h (2 mile/h). If rows of injections were 20 cm (8 in) apart, the application rate would be equivalent to 2300 L/ha (246 gal/acre). The proper amount of chlorpyrifos was mixed with water in the tank to comply with recommendations by the manufacturer (1.2 kg a.i./ha). This mixture was applied in four replicated plots. The plots were immediately irrigated for two hours, equal to 1.3 cm (1/2 in) of precipitation.   Fifteen injection holes per plot were selected and five samples surrounding each injection hole were taken for residue analysis. One of the sample locations was the point of injection. Two of the locations were 3.8 cm and 7.6 cm away from the injection point at a 30° angle from the line perpendicular to the line of travel (northeast direction). The other two locations were 3.8 and 7.6 cm away from the point of injection on a line opposite to the line of travel (south direction). Each of the 75 samples taken from these five locations was subdivided into three subsamples representing thatch, the first 2.5 cm (1 in) of soil, and the second 2.5 cm of soil. The thatch and soil increments were pooled for each depth increment, placed in separate plastic bags for each replication, and frozen at -18°C (0°F) until analyzed.

Another complete set of samples as explained previously was taken on June 20 to determine the horizontal and vertical movement of chlorpyrifos seven days after its application.   Between the first sampling and second sampling, the plots received additional irrigation; 13 mm (1/2 in) on June 15, 6 mm (1/4 in) on June 20, and 3 mm (0.13 in) rain was recorded on June 17. Two subsamples from each of the pooled sample bags were taken and analyzed for insecticide residues using a Hewlett Packard Model 5890 gas chromatograph.

There was negligible damage to the turf by the injection wheel. It was difficult to find the holes produced by the wheel immediately after application. Analysis of samples revealed that the injector wheel consistently delivered the desired amount of insecticide at the point of injection. The residue levels in samples taken both 1 and 7 days after application showed that there was little lateral movement of insecticide from the point of injection at all three depths. Chlorpyrifos concentration in samples taken 1 day after application decreased rapidly with depth. At the point of injection, the mean concentration decreased from 226 to 61 to 2 ppm for the top to bottom sample locations. An even greater downward reduction trend in insecticide concentrations from top to bottom samples occurred at 7 days after application. These data indicated little vertical movement of chlorpyrifos in soil.

The residue analyses indicate that chlorpyrifos moves very little in the soil when injected with a point injector wheel. Also, due to considerable concern about pesticide contamination of ground water, it is likely that pesticide manufacturers will attempt to minimize movement of their pesticides in soil. Little is known about the specific movement of Japanese beetle (Papillia japonica Newman) grubs, but it is generally believed that they move little if food and other conditions are satisfactory. It is likely that their movement is not sufficient for most of the grubs to encounter a lethal dose of an insecticide applied with point injectors unless the points of injection are very closely spaced.

A further concern is the potential complications that may arise when mounting several of these wheels on a common toolbar with a practical width for treating large areas. Based on our results with chlorpyrifos, the wheels should be close together. Such close spacing would be expensive and may be impossible with known injector wheels. Unless the grub or the insecticide moves considerably, a system to distribute insecticides with considerably less variation than distribution from the point injector wheel is needed to efficiently and effectively control turfgrass grubs.

Slit-injection

Two additional systems for subsurface placement of Dursban 4E (chlorpyrifos) and Triumph 4E (isazophos) insecticides in turf were evaluated. First, a high pressure injection spray unit (the HPI-2000), manufactured by the Cross Equipment Co., Albany, GA, was used to apply two test insecticides. The unit consisted of a skid-mounted boom (Figure 3) with 18 nozzles at intervals of 7.2 cm (3 in). The insecticides were injected directly into the thatch as a solid stream at a pressure of 12,400 kPa (1800 psi).

The third injection equipment used was the RainSaver Jr. (RS), manufactured by Clearwater Industries, Nezperce, ID. It has thin coulter

Figure 3   The RainSaver Jr. slit-injection system

Figure 4   High pressure injection system

blades mounted 7.2 cm (3 in) apart on a common shaft to open narrow slits in turf. A series of knives immediately follow the coulters and widen the slits. Liquid insecticide is dribbled into these slits at atmospheric pressure (Figure 4). The coulters were adjusted so the insecticide was placed at a depth of about 1.9 cm (3/4 in) below the sod surface.

The HPI and RS systems were compared to the conventional method of applying liquid insecticides for grub control, namely, a broadcast spray followed by irrigation. For the conventionally treated plots, Triumph 4E and Dursban 4E, were applied at 2.24 kg active ingredients/hectare (2 lb a.i/acre) to a golf course fairway with no thatch at Canal Fulton, Ohio, and one at Wooster, Ohio, having 1.3 to 2.5 cm (0.5 to 1 in) of dense thatch. Both insecticides were applied at a volume of 1628 liter of spray mixture per hectare (174 gallons/acre) and the treatment at Canal Fulton was immediately irrigated with 0.6 cm (1/4 in) water.

Residue Analysis

To compare residues remaining on the grass blades following treatment, samples of grass were collected from each treatment on the day of application. Plots receiving broadcast sprays were sampled before and after irrigation, while those treated with the HPI and RS systems were sampled immediately after application. Residues were extracted within three hours after collection and analyzed by gas chromatography (Hewlett Packard Model 5890) within three days.

Results of the residue analyses are shown in Table 1. Irrigation following the conventional application reduced grass residues of Triumph by 43% (from 47.3 to 26.8 ppm) and Dursban by 51% (from 78.8 to 38.8 ppm). Residues from both injection methods were much less than that from the conventional spraying followed by irrigation. HPI treatments reduced Triumph residues by 62% (from 26.8 to 10.1 ppm), and Dursban residues by 38% (from 38.8 to 24 ppm) from residues remaining after irrigation of plots treated with conventional spraying at Canal Fulton location. The RS treatments resulted in the lowest grass residues. Residues of both Triumph and Dursban were reduced to less than 1.4 ppm, a reduction of nearly 95% when compared with conventional sprays followed by irrigation.

Grub Control

Effectiveness of the treatments for control of Japanese beetle grubs was determined 34 days following treatment. The results are presented in Table 2. Although both injection systems provided better control than the conventional system when using Dursban, generally the control from application of Dursban with the HPI and conventional systems in turf with and without thatch was less than expected. Control with Dursban and the RS injection method, however, was considerably greater than control with the conventional

TABLE 1 -- Insecticide Residues on Grass Blades following Conventional and Injection Application Methods - 1988

Insecticide Used	Method of Application	Mean (± S.E.M.) Residues (ppm)	
		Canal Fulton[1] (no thatch)	Wooster[2] (1 in thatch)
Triumph 4E	Conventional spray	26.8 ± 3.91*	--
Triumph 4E	HPI-2000	10.1 ± 0.43	13.2 ± 2.24
Triumph 4E	RainSaver Jr.	1.4 ± 0.31	1.3 ± 0.08
Dursban 4E	Conventional spray	38.8 ± 1.79*	--
Dursban 4E	HPI-2000	24.0 ± 2.71	11.1 ± 0.77
Dursban 4E	RainSaver Jr.	1.3 ± 0.26	1.5 ± 0.29

*Residues after irrigation. Residues measured before irrigation for Triumph and Dursban were 47.3 ± 4.33 and 78.8 ± 8.72 ppm, respectively.
[1]Mean residues based on separate GC analysis of 5 pooled 1 ft^2 area samples of grass blades from each of 3 replicates.
[2]Mean residues based on separate GC analysis of 8 pooled 1 ft^2 area samples of grass blades from each of 2 replicates.

TABLE 2 -- Mean Percent Control of Japanese Beetle Larvae using Three Application Techniques - 1988

Application Method	Insecticide	Mean % Control*	
		Canal Fulton (no thatch)	Wooster (1 in thatch)
Conventional Spray	Triumph 4E	87 d	76 d
"	Dursban 4E	26 b	5 ab
HPI-2000	Triumph 4E	82 d	57 bcd
"	Dursban 4E	40 bc	21 abcd
RainSaver (RS)	Triumph 4E	82 d	69 cd
"	Dursban 4E	72 cd	48 abcd
	Untreated	0 a	0 a

* Mean of 3 replicates. Means in a column followed by the same letter are not significantly different according to Duncan's New Multiple Range Test at $\underline{P}$ = 0.05.

method (5% vs 48%) in turf with thatch. With no thatch, the conventional system provided 26% control whereas 72% control was achieved with the RS.

With Triumph 4E, all application methods provided satisfactory control in grass with no thatch. The conventional system provided slightly better control than that from the injection methods. In grass with thatch, the conventional method again outperformed the injection methods when using Triumph 4E. Between the two injection systems, the RS provided slightly better control than the HPI system.

Generally, grub control with HPI was poor in grass with thatch because of inadequate penetration of insecticides into the zone of grub habitation. Control with RS was better but not as good as that from the conventional application of Triumph.

SUMMARY AND CONCLUSION

Results from this study indicate that insecticides can be successfully injected with little damage to turf. Among the three injection systems investigated (point injection, high pressure injection, and coulter/knife injection), the RainSaver Jr. system which uses a combination coulter and knife assembly showed greater potential for grub control in turf with insecticides than the other two. Poor distribution of the insecticide was the main problem with the injector wheel. The HPI system may have limited application in cool-season areas where thatches are often too dense to permit penetration of the insecticides to the soil level. However, this system may be adequate for applying insecticides to control insects such as mole crickets in warm season areas where thatch may be less dense.

With better depth control, these systems should more accurately deliver materials directly and only to the subsurface target zone and allow rates much less than those currently needed for grub control. In addition, other materials, now ineffective because of immobility in thatch and/or soil, may show effectiveness when placed directly in the zone of pest activity.

The placement of insecticides in the soil under turfgrass without thatch, or at the soil surface under thatch, may increase the potential for these materials to move downward through the soil. However, this potential should be substantially reduced, if not eliminated, by greatly reduced application rates.

Potentially, the subsurface placement could be advantageously used with many materials, such as certain fungicides, herbicides, and nematodes parasitic to insects. Further research on subsurface placement of insecticides is planned for 1990-91. In addition to more tests with liquid insecticides, several pieces of overseeding equipment with coulters will be modified and evaluated for their potential to place granular products beneath the turf.

ACKNOWLEDGEMENTS

The authors thank Dr. J. L. Baker, Iowa State University and Ed Darlington, Liquid Ed Inc., Florida, for making point injector wheels available to us; Clearwater Industries, Nezperce, ID, and Cross Equipment Co., Inc., Georgia, AL, for providing injection equipment; D. L. Collins, J. Mason, and K. T. Power for their technical assistance; J. Doty and R. Clason for their assistance in the preparation of this manuscript.

REFERENCES

[1]   Edwards, C.A. and Adams, R.S., "Persistent Pesticides in the Environment," In Critical Reviews in Environmental Control, Vol 1, Issue 1, Cleveland, Ohio, 1970.

[2]   Edwards, C.A., "Actions and Fate of Organic Chemicals in Soil: Insecticides," In C.A. I. Goring & J.W. Hamaker [eds], Organic Chemicals in the Soil Environment, M. Dekker, N.Y., Vol. 1, pp 4340, and Vol. 2, 1972, pp 503.

[3]   Harris, C.R., "Factors Influencing the Effectiveness of Soil Insecticides," Annual Review of Entomology, Vol. 17, 1972, pp 177-198.

[4]   Sears, M.K. and Chapman, R.A., "Persistence and Movement of Four Insecticides Applied to Turfgrass," Journal of Economic Entomology, Vol. 72, No 2, 1979, pp 272-274.

[5]   Kuhr, R.J. and Tashiro, H., "Distribution and Persistence of Chlorpyrifos and Diazinon Applied to Turf," Bulletin of Environmental Contamination Toxicology, Vol. 20, 1978, pp 652-656.

[6]   Niemczyk, H.D., and Krueger, H.R., "Binding of Insecticides on Turfgrass Thatch," Advances in Turfgrass Entomology, 1982, Chemlawn Corp., Worthington, Ohio, pp 61-63.

[7]   Niemczyk, H.D. and Krueger, H.R., "Persistence and Mobility of Isazofos in Turfgrass Thatch and Soil," Journal of Economic Entomology, Vol 80. No 4, 1987, pp 950-952.

[8]   Branham, B.E. and Wehner, D.J., "The Fate of Diazinon Applied to Thatched Turf," Agronomy Journal, Vol. 10, 1985, pp 101-104.

[9]   Anonymous, Hearing Before FIFRA Scientific Advisory Panel on the Cancellation of Diazinon for use on Golf Courses and Sod Farms, Environmental Protection Agency, Washington, D.C., May 21, 1986.

[10]   Baker, J.L. and Timmons, D.R., "Effect of Method Timing and Number of Fertilizer Applications on Corn Yield and Nutrient Uptake," Agronomy Abstracts, 1984, Iowa State University, Ames, IA, pp 197.

[11]   Baker, J.L., Timmons, D.R. and Cruse, R.M., "A Comparison of Applying Fluid Fertilizers in No-Tillage Corn," Paper presented at Fluid Fertilizer Foundation Symposium, March 14-15, South Padre Island, TX, 1985.

[12]   Baker, J.L., Colvin, T.S., Marley, S.G. and Dawelbeit, M., "Improved Fertilizer Management with a Point Injector Applicator," ASAE Paper 85-1516, ASAE, St. Joseph, MI, 1985.

[13]   Ozkan, H.E., Reichard, D.L., Niemczyk, H.D., Klein, M.G., and Krueger, H.R., "A Subsurface Point Injector Applicator for Turfgrass Insecticides," Applied Engineering in Agriculture, Vol. 6, No 1, 1990, pp 5-8.

Arthur G. Appel and Larry G. Woody

INDIVIDUAL MOUND TREATMENT FOR RAPID CONTROL OF FIRE ANTS

REFERENCE: Appel, A. G. and Woody, L. G.,
"Individual Mound Treatment For Rapid Control Of
Fire Ants", Pesticide Formulations and
Application Systems: 10th Volume, ASTM STP 1078,
L. E. Bode, J. L. Hazen, and D. G. Chasin, Eds.,
American Society for Testing and Materials,
Philadelphia, 1990.

ABSTRACT: Techniques for controlling individual
mounds of the red and black imported fire ants
were evaluated. Of the five techniques the
following combination provided the best control:
perimeter spray followed by disturbing the mound,
applying the spray topically, and injecting the
spray into at least 12 locations. Mounds were
inactive within 3 minutes. This procedure used
approximately 2 cans of wasp spray per mound and
usually killed the mound. Approximately 5% of
the mounds either contained live ants or
relocated within 24 hours after treatment.
Within 7 days, the relocated mounds had died, and
there was no further relocation after 30 days.
This procedure controlled ants effectively
rapidly and killed treated mounds in most cases.

KEYWORDS: fire ant, insecticide treatment, mound
application, mound injection, control

The red and black imported fire ants, Solenopsis
invicta Buren and S. richteri Forel (Hymenoptera:
Formicidae), are distributed throughout much of the
southeastern United States from central Texas through the
southern parts of North Carolina. Fire ants are serious
agricultural and urban pests that readily sting and bite.
The sting causes a painful burning sensation, then the area
reddens and the area swells into a wheal and a pustule
forms within a day [1]. These ants nest in large colonies
forming above-ground mounds that may be located in or next
to vertical objects such as trees and fence posts and
electrical, telephone, and television equipment.

Dr. Appel is an assistant professor in the Department
of Entomology and Alabama Agricultural Experiment Station,
Auburn University, AL 36849-5413; Mr. Woody is Staff
Manager Product Evaluation and Selection, BellSouth
Services, Birmingham, AL 35283.

Disruption of service and routine maintenance often
necessitates entering and working in infested equipment.
Rapid knock down and kill of the ants would allow safe
access into the equipment. The purpose of this study was
to develop a rapid and effective technique for immediate
control of fire ants.

MATERIALS AND METHODS

     A fire ant (S. invicta) infested Christmas tree farm
located approximately 10 km from the Auburn University
campus in Lee County, Alabama was selected as the field
site. Thirty mounds, 40 to 90 cm in diameter and
containing over 50,000 ants, were selected for treatment
September 28, 1988. These were large mature late season
mounds; most mounds are smaller than those used in this
study.

     Whitmire Wasp Stopper CF (0.10% d-trans Allethrin)
(Rainbow Waspkiller II) aerosol spray formulation was
applied to the mounds. This formulation was selected
because it provides rapid knockdown and does not damage the
plastics used in telecommunications equipment [2]. The
following application methods were employed: topical
application alone (approximately 1, 14 oz can), application
of 1 can using an injector rod only (no topical
application) (Fig. 1a), approximately 8 insertions of 2 sec
spraying/insertion, topical and rod application [25% of 1
can topical, 75% rod (approximately 6 insertions of 2 sec
spraying/insertion)], circle mound with a barrier spray
(Fig. 1b) followed by topical and rod application
(approximately 6 insertions of 2 sec spraying/insertion),
and circle mound with a barrier spray, break up mound (Fig.
1c) followed by topical (Fig 1d) and rod application
(approximately 6 insertions of 2 sec spraying/insertion).
Each treatment was replicated 6 times.

     Performance was evaluated 3 min after application, 1,
7, and 30 days after treatment. Evaluations consisted of
tapping and probing mounds to determine the presence of
live ants. Repellency of the spray was evaluated by noting
the presence of new mounds within a 1 m radius of the
treated mound. These mounds were considered relocations.
Six untreated control mounds were also selected and
evaluated as above. Data were analyzed with chi-square
(X^2) tests.

RESULTS AND DISCUSSION

     There was no mound mortality or relocation with the
untreated control mounds. Topical and rod treatments alone
did not give 100% kill 3 min after treatment (Table 1).
All combined treatments of topical and rod application (n =
18), however, gave 100% kill within 3 min. Topical
treatments resulted in only 17% mound kill at 7 and 30 days

combined applications (topical and rodding) gave 100% mound kill after 3 min and at 1, 7, and 30 days after treatment. Therefore combined applications were significantly ($p < 0.05$) more efficacious more quickly than single method applications.

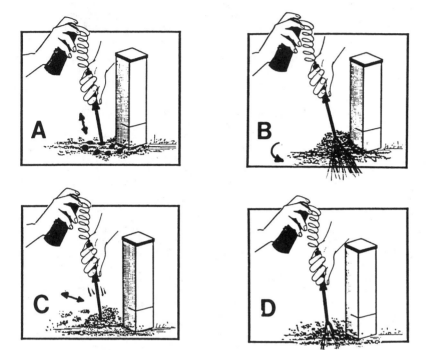

FIG. 1 -- Treatment of individual mounds with an injector rod for fire ant control; a, rodding, b, barrier treatment, c, breaking up the mound, d, topical application.

Table 1 -- Performance of different application techniques for control of red imported fire ant mounds with Wasp Stopper CF.

Treatment	n	% killed (K) and relocated (R) at day							
		0		1		7		30	
		K	R	K	R	K	R	K	R
Control	6	0	--	0	0	0	0	0	0
Topical	6	0	--	0	17	17	17	17	33
Rod	6	0	--	0	33	100	33	100	33
Topical+Rod	6	100	--	100	33	100	33	100	33
Circle+ Topical+Rod	6	100	--	100	17	100	17	100	17
Circle+Breakup+ Topical+Rod	6	100	--	100	17	100	0	100	0

A maximum of 33% of the treated mounds relocated (Table 1). More relocations occurred in topical and rodding methods alone and the combined topical and rodding method than with treatments including an initial perimeter spray. This may indicate that initial relocations occur very quickly during treatment and that the perimeter spray repels the relocating ants. In all cases, however, the relocated mounds were not abutted to the treated mounds.

In conclusion, this aerosol spray formulation is effective against the red imported fire ant when properly applied. The most effective technique is to apply a perimeter spray followed by breaking up the mound, topically spraying on the ants, and then rodding the spray into the mound. This technique gives 100% kill within 3 min and limits mound relocations to ≤ 17%. This application technique also results in 100% mound kill over at least a 30 day period.

ACKNOWLEDGMENTS

The authors thank Jeff Steeley, Rainbow Manufacturing, Birmingham, AL and Whitmire Research Laboratories, St. Louis, MO for their support of this research. Alabama Agricultural Experiment Station Journal Series No. 17-902439P.

REFERENCES

[1] Vinson, S. V. and Sorensen, A. A., Imported Fire Ants: Life History and Impact, Booklet, Texas Department of Agriculture, 1986.

[2] Appel, A. G. and Woody, L. G., "Test methods for pest control in the telecommunications industry", Pesticide Formulations and Application Systems: 8th Volume, ASTM STP 980, D. A. Hovde and G. B. Beestman Eds., American Society for Testing and Materials, Philadelphia, 1988.

Author Index

A

Adams, A. J., 156
Akesson, N. B., 170
Appel, A. G., 248

C

Chapple, A. C., 156
Curcio, L. N., 11

D

Dailey, O. D., Jr., 26
Doane, W. M., 17
Dowler, C. C., 26

F

Funk, R. C., 111

G

Gaultney, L. D., 126
Gibbs, R. E., 170
Glaze, N. C., 26
Gottfried, S., 226
Group, E. F., Jr., 71

H

Hall, F. R., 156, 184
Hart, W. E., 126
Herrera, J., 226
Hummel, J. W., 111

K

Kemble, L. J., 226
Klein, M. G., 236
Knapp, J. L., 84
Krishnan, P., 226
Krueger, H. R., 236

L

Leedahl, A. O., 140
Lehtinen, J., 184
Luttrell, R. G., 56

M

Manthey, F. A., 71
Moechnig, B. W., 38

N

Nalewaja, J. D., 71
Niemczyk, H. D., 236

O

Ozkan, H. E., 236

R

Rachman, N. J., 3
Raymond, M. L., 42
Reed, J. P., 184
Reichard, D. L., 184, 236
Riedel, R. M., 184
Riley, C. M., 204

Subject Index

A

Adjuvants, 71, 93
Agglomeration, 38
Airborne dust measurements, 42
Application rate control, 111,
    126, 140
    drift, 156, 142, 170, 184, 204
Assessment, risk, 3, 11
Assimilation, carbon dioxide, 84
Atomizer, 84, 170, 142, 184
Atrazine, 26
Avian toxicity, 38

B

Birch, white, 93
Boom tube, 140

C

Calibration, multivariate, 111
Canopy partitioning, 184
Carriers, 71
    dust measurements, 42
    granular, 38, 42
    inert, 42
Chemigation, 26
Contamination
    groundwater, 17, 26
    reduced, 140
Corn, 71
Cyclodextrin, 26

D

Deposit density, 56
Dietary risk level, 3
Diffuse reflectance, 111
Direct injection, 126
Dosage, spray deposit, 56
Drift, 156, 184, 204
    modeling, 170
    potential, 142
    spray, loss, 170

Droplet drying, 93
Droplet frequency analyzer, 170
Droplet size, 56, 142, 156, 170
Droplet spreading, 93
Dry flowable pesticides, 126
Dust, airborne, carrier, 42

E

Efficacy, pesticide, 184
    cyclodextrin, 26
    insecticide spray, 56
Encapsulation, 17
Environmental Protection
    Agency, 11
Environmental risk, 3, 11

F

Fertilizer
    dry granular, 140
    inert granular carrier, 38
Fire ants, 248
Flat plate collectors, 204
Florida beggarweed, 26
Frequency analyzer, 170
Fungicide, inert granular
    carrier, 38

G

Gelatinization, starch, 17
Glyphosate, 93
Granular carrier, 38, 42
Granular formulation, for
    sprayer, 126
Gravimetric dust
    measurements, 42
Green foxtail, 71
Groundwater contamination,
    17, 26
Grub control, 236

255